HOW TO ENJOY
YOUR WEEDS

HOW TO ENJOY
YOUR WEEDS

by

Audrey Wynne Hatfield
F.R.H.S.

FREDERICK MULLER

*First published in Great Britain 1969
by Frederick Muller Ltd, Fleet Street, London, E.C.4*

Copyright © 1969 by Audrey Wynne Hatfield

*Printed and bound by
Cox & Wyman Ltd
London, Fakenham and Reading*

For

Joan E. B. H. Biggin

Introduction

To enjoy your weeds you must know and exploit them. Living isolated in the midst of the country, surrounded by fields, woods and coppices, I am very familiar with weeds; I have most of them and nearly all the wild plants of Hertfordshire. My soil must harbour great quantities of dormant seeds, as every patch of ground, after being cleared for some desired planting, is almost immediately covered not only with newly awakened seedlings, but with eager colonisers from the verge and hedges. Either the original flora, and the seed-carrying fauna, resent my intrusion, or their offspring are agog to come into my comparatively less competitive acre. My simple choice has been to get to know them and deal with them according to their worth.

The plants we call 'weeds' are either long-established wildings or they may be survivors from past deliberate introductions, which are invading gardens where the owners do not want them. But, strange as it may appear to some harassed gardeners, most of their hated weeds have good uses. Some make health-giving food or drink, and were in fact encouraged by our forefathers as of prime importance in their much-restricted diet and medicines, and it should be remembered that they still provide the basic, well-tried remedies of the homeopath and the herbalist. Like all herbage, weeds have now been found to possess peculiar qualities which, when understood, can be made to contribute a great deal in various ways to help the robust growth of the plants we wish to cultivate. All weeds are interesting in some aspect.

I am not an advocate for weedy gardens, but I am concerned with knowledgeable and enjoyable methods of coping with them. They are every gardener's concern as no soil of any worth could be free from weeds. The vitality of some of their seeds can endure for a long period of time within the ground where conditions are right for their dormant preservation, and where, surrounded by an atmosphere rightly charged with carbon dioxide, they can sleep until the earth is disturbed and they are brought to the surface to be awakened by light and fresh air. A famous instance of such a revival was the spectacular display of scarlet poppies arising from the newly dug trenches, the shell-holes and graves on the battlefields in Flanders during the 1914–18 War.

New weeds and fresh supplies of old ones continually arrive by varied means; to country or town gardens they come on currents of air; they arrive on, or inside, migratory or native birds, or animals, and any waste ground in a blackened industrial area is mysteriously sown with a collection of struggling herbage. Seeds come from near and far in packing, in mud on transport vehicles, and in cargoes; districts around our ports are sometimes enlivened with foreign plants, most of which die after a season's flowering, but a few may survive as they have done in the past. A surprising number of seeds are carried on clothing. To prove this possibility, after a walk in the country scrape a shoe or trouser leg over a clean sheet of paper, empty the dust on to a tray that is spread with a layer of damped sterile peat, sterilised soil or sand, and see what germinates! Plants have been carried in this way for many thousands of years, even as seeds on the garments and packages of the earliest, primitive men and subsequent travellers.

New plants from distant countries would spring up along the tracks of early nomadic tribes; more seeds would come inside the bags of grain and on cattle brought here by immigrant farmers of the New Stone Age, the Bronze Age, the early Iron Age, and with later settlers and invaders. And many of these plants were then known to be good food. There was a sensational example of this in 1950 when botanists working with archaeologists were able to give a detailed description of the Tollund Man's last meal. Some corpses were recovered from early Iron Age burial grounds in the Danish raised bogs at Tollund, where they had lain for some two thousand years. One body was that of a chieftain who, considering that he had been hanged, appeared benign and serene. He was so perfectly preserved in the peat, save for a bent nose, that botanists were able to analyse his stomach contents and found the remains of his last meal, which had consisted of a gruel made from roughly ground seeds. There was a large proportion of linseed and barley mixed with smaller amounts of 'weeds', such wild plants as pale persicaria, black bindweed, gold of pleasure, fat hen, hemp nettle, wild pansy and corn spurrey. This gruel would be more nutritious than our breakfast cereals, as it was rich in proteins and vitamins. The pansy seeds would give it a mildly stimulating effect for the heart and respiratory organs; they were also a gentle laxative. Ample evidence was discovered to prove that all these plants were deliberately collected or cultivated by those people. As most of them were also our native wild plants, they would be encouraged here and consumed in the same way. Corn spurrey was actually introduced into Britain, with linseed, or flax, at that time in the early Iron Age, and it was grown as a crop for its oily seeds in Roman times and for centuries later.

During the three hundred years of the busy comings and goings of the Roman occupation our native flora was enriched by quantities of exotic seeds that were brought to Britain from all over Europe and Asia. Some were deliberate introductions to be cultivated as customary Latin food and medicines, and some to provide the cosmetics and the floral decorations that were necessary adornments of sophisticated Roman living. Other seeds would be accidentally spread about from the packings of imports, and from the clothing and equipment of the much-travelled legions who built and used the new roads, nearly seven thousand miles of them from end to end of Britain. An early observer, Dr John Dee, who lived during the reign of Queen Elizabeth I, recorded his finding the medicinal herbs planted by the Romans still growing along Hadrian's wall. As, when we know our plants, we find not only Roman, but numerous older and some more recent romantic introductions still flourishing as weeds in our gardens. Any weed is more enjoyable when we know something interesting about its associations and its ancient uses.

Nowadays we have abundant evidence of the periods when certain plants existed in different areas of Great Britain; their fossil remains, pollen and seeds, are to be found in various strata of our land's formation. Many excavations have yielded important discoveries, but the most puzzling are the rare instances when a few isolated remains of some particular plant have turned up in the levels of a period long before there has been any further evidence of the plant's existence in this country. For instance, who brought the lettuce that grew in the levels of the Mesolithic settlement on the Isle of Wight, where six perfect seeds were found? It was *Lactuca sativa*, a type native to South-west Asia or Siberia, that later, millenniums later, was introduced into Europe and became our salad plant. Unlike our native wild lettuces, this species would have been too tender to survive unaided, and multiply to become one of the weeds we find difficult to control.

Our garden weeds, from small annuals to greedy perennials, are, in general, plants that have persisted through the great opposition of the earth's changes and catastrophes, at times being mercilessly shifted about by quakes, glacial periods, floods and avalanches, enduring periods of intense cold, or heat, meanwhile becoming less sensitive to types of soil and climate, and evolving those aggressive methods of defence that defeat our efforts but must evoke our interest and respect. A few of these survivors became self-fertilising, relying neither on insects nor wind; some developed roots that could run underground like trains, thrusting through such solid obstacles as our potatoes, and sending up shoots to the surface at short, or long, intervals to become independent, menacing plants. Others travelled above ground, rooting as they progressed, and forming their complicated networks of abundant colonies.

Certain of our weeds developed a cunning capacity for staggering the germination timing of the weeds they would simultaneously ripen. A number of these fruits could germinate almost immediately they touched the ground, others were delayed for an interval of time, while the rest remained dormant for months. This assured the chances that some offspring would survive if others encountered impossible weather, the gardener's hoe, or other ungenial conditions.

Weeds of all kinds appear to have evolved marvellous tricks to tease and embarrass gardeners; they are by nature the determined plants that have been the most tolerant and are now the most successful. We could never for long be free of them and we would be the poorer without them.

A good garden soil should contain quantities of the nutrient or 'fertilising' elements, nitrogen, phosphorus and potassium, with smaller amounts of magnesium and sulphur predominating over the minor, or 'trace', elements such as iron, zinc, manganese, copper, cobalt, boron, the rare base metal molybdenum, with other things according to the rock foundation and the mineral content of the site. These are naturally sought and assimilated by plants as they are necessary for their well-being. They are also essential for ours, and we can get them only directly or indirectly through our edible crops or the varied pastures where cattle graze. If these elements are absent from our garden soil or if they are rendered unavailable through too high a lime content, we, and our plants, suffer from their deficiency. Some plants choose and take up a greater amount of one element than do others. For instance, we value spinach for its iron and vitamin content, but some weeds are much richer in these and other essentials. So it follows that when we 'weed' our gardens we take out plants that have robbed the soil and have not only stored some of its vital components but have manufactured certain valuable secretions. If we take away either a weed, which may be a miser of some valuable element, or the waste from a harvested vegetable, and if we burn it on a bonfire we seriously deplete our soil and rob desirable plants not only of their elemental requirements but also of humus.

A gardener's responsibility is to return everything he takes from his ground. Bonfires should be kindled merely to burn such things as woody prunings, old stakes, pea and bean sticks, with dollops of obstinate clay which, when burnt, becomes red and friable, rich in mineral salts, and, with the wood ash, makes an excellent addition to the soil, *except* near mint, which is badly affected by any bonfire ash.

Everything that is taken from the soil goes back to it, and with interest, by composting. Every weed, haulm, leaf and the lawn mowings, with kitchen waste, torn-up paper (not greaseproof), vacuum-cleaner dust and fluff, feathers, hay, straw and any available waste from a fishmonger's or greengrocer's shop should go into the compost heap. This collection becomes one of the pleasures of gardening, rather like preparing tasty banquets, with weeds providing the 'seasoning' and valuable nutrients.

An efficient heap should measure about 6 ft by 4 ft, and it must sufficiently heat up to consume weed seeds and their most persistent roots that are a prime source of indispensable plant food. The sides of the heap should be enclosed to prevent currents of air keeping the margins cool. This can be achieved in several ways: by stacking turves grass-side down to make the side walls around the site. By making wooden sides with planks fixed to upright supports and arranging a movable end so that the composted stuff can be easily removed. An efficient cage-like structure is made with wire-netting fastened to stakes driven into the ground, leaving one end hinged to open, or separate so that it may be lifted away. To retain the heat that is generated, this light open pen must be lined with straw, or with sheets of newspaper which are turned into the compost as they rot, and are replaced with sound ones. However it is constructed, the heap must stand on bare earth which admits the necessary soil bacteria to enter the waste matter and do its job of breaking down and decomposing.

The materials to be composted should be placed in layers about 6 in. thick, sandwiched with thinner layers of soil, garnished with a few sprays of aromatic herbs, and a sprinkling of a properly balanced and blended herbal activator, or one made from seaweed, the weed rich in all the basic elements which are washed into the sea from river beds. Either of these is far superior to chemicals that are sold for the purpose.

When the heap reaches 3 or 4 ft in height the contents must be covered and sealed with soil like a potato clamp. All the materials will soon rot down to a rich humus containing most of the soil's requirements, and it should look appetising as dark, crumbly Christmas pudding.

It must be appreciated that no plants are fit for the compost heap if they have been poisoned or sprayed with a chemical pesticide. And grass is not eligible if it has been mown from a lawn that has been given any selective weed-killer. But if lawn-sand has been used the mowings are good.

If the compost heap can be made under an elderberry bush or a birch tree, the excretions from their roots, together with their fallen leaves, will assist fermentation, making the compost light and especially effective for restoring the soil.

Well-rotted compost is of greatest value as a top or near top dressing that will be gradually washed down to the plant roots, but it is too precious to bury deeply when it will go down too far for them to reach. When digging a bed for planting, any weeds, excluding the live roots of perennials, should be put in nearly a foot deep and chopped with a sharp spade, and covered with uncomposted kitchen waste, then soil and compost on top. I have had spectacular results with this method; the plant roots seem to hurry down to the egg-shells, tea-leaves, peel, fish, vegetable waste and weeds, producing strong, healthy top-growth as they gorge.

The value of weeds in the compost heap and back into the soil from whence they came is probably even greater than we yet realise, as new evidence of the power of plants is being continually observed. An exciting discovery of recent years was made by a Dutch nurseryman who planted African marigolds of the variety Colorado Sunshine (*Tagetes erecta*), as a cut flower crop after removing his daffodil bulbs. Finding that the marigolds had destroyed his nematode soil pest, the narcissus eelworm, he passed the information to a research station. There experiments were started with various kinds of Tagetes. The nurseryman had happened on a lucky choice as the hybrids he might have chosen for cut blooms gave no such pesticide results, but it was found that both the so-called African, and the French marigold (*Tagetes patula*), which are really Mexican wild plants, could kill other plant-destroying nematodes too, and at a range of 3 ft, but the beneficial types of eel worms that do not feed on healthy roots were not affected.

The details of more experiments with various Tagetes species make too long a story for this space, but an English research association pursuing their tests with *Tagetes minuta*, direct from its native Mexico, found this plant not only dealt with all kinds of destructive eel worms and dismissed wireworms, millepedes and various root-eating pests from its vicinity, but in many instances it also killed couch grass, convolvulus, ground ivy, ground elder, horsetail and other persistent weeds that defy most poisons. The lethal action worked only on starchy roots and had no effect on woody ones like roses, fruit bushes and shrubs. Where it had grown the soil was enriched as well as cleansed, its texture was refined and lumps of clay were broken up.

TAGETES MINUTA

According to archaeological findings, this Mexican weed's powers were no new discovery. Its flower umbels have been identified as those that were pictured accompanying the various crops which were painted on the vases buried with Chimu farmers and gardeners of ancient Equador, and those of the Chavins from pre-Inca Peru, the civilisations which thousands of years ago bred potatoes, tomatoes, runner and French beans, maize and other crops we also value. This was the plant sacred to their gods of agriculture which enabled those people, who had very restricted areas of usable land, to grow continually and successfully for centuries on the same ground such plants as shared the same eel worm risks with each other. We could not grow such crops on the same ground for more than two years without inviting trouble. Yet some of those areas were cultivated without rest or rotation for two thousand years, with sea-bird guano, fish waste and Tagetes. Without this protective herb the cultivated plants would certainly have soon become the victims of eel worms and perished. To those farmers without equipment to see the pests, their crop failures would appear to be the curse of their gods.

The story of this Mexican marigold exemplifies various aspects of most plants. Its rapid and wide distribution as a weed has been recent enough to be followed. It is a lusty annual of giant size, with umbels of small, cream, disc-like flowers, and the seeds were sent from Mexico to an Australian park, to be grown as a rare plant with decorative foliage. It was so happy there that its seeds ripened and by all the methods of transport seeds can use they travelled over Australia to become a common weed. During the 1914–18 War some seeds accidentally arrived in South Africa in the fodder for the horses of the Australian artillery. This marigold liked that country too, and romped away as a flourishing weed. In Rhodesia the powerful qualities of the secretions and excretions of this now common weed provide a cheap and effective answer to some tobacco growers' problems, as it attacks their soil pests and other weeds, and the liquid from the leaves after they have been boiled makes a very efficient pesticide spray. In this aspect the Tagetes is an outstanding example of all plants' ability to manufacture their own brand of secretion that is characteristic of their race, their family, though the members, the species, vary considerably in their accumulations. Plants of the genus Tagetes are members of the *Compositae* family, the daisy tribe, and while all daisy-flowered plants may produce much the same substances in very varying degrees, *Tagetes minuta* by far outdoes them all in quantity and potency. But the same natural pesticide compound is made by its relation, the wild pyrethrum of Africa, and this has been available for many years as about the safest of all pesticides that are commercially exploited.

It is obvious that the compounds which plants excrete through their roots have a direct influence on the vitality of the soil; some stimulate the growth of valuable organisms, others poison the necessary bacteria. Perhaps less obvious is the effect of their exhalations, their scents of volatile substances which make them acceptable or repellent to other types. A noticeable example is the ethylene gas breathed out by some ripening fruits, especially apples, bananas, tomatoes, peaches, avocados; when in a room near such fruits, daffodils and carnations can be seen to suffer from the gas and roses are sickened by it to a lesser degree. Ethylene is also generated by damaged leaves of green vegetables in the garden to the annoyance of growing plants and flowers near by.

Another instance of the reality of these substances is the particular breath of violets that affects human beings by so numbing the nerves which control our sense of smell that gardeners who grow a lot of these flowers are unable to enjoy their fragrance.

With all these influences it is obvious that plants must be great individualists, and anyone who has tended a garden for many seasons will have become aware that they have their affinities and their enemies among other plants. Some will thrive in one place but will not even try in another. This is often because their bedfellows are either sympathetic and encouraging, or they may be detrimental and stunting. Few of us can boast of an herbaceous border where some plants have not appeared sadly defeated even though their soil requirements have been supplied, and many a gardener has been puzzled by some failure in his vegetable plot. These disappointments can be caused by the lack of suitable influences or the presence of aversions, and to avoid them we should aim at the knowledgeable placing together of sympathetic plants.

I have experienced unmistakable instances of this plant symbiosis, and I have heard of and read about others, which I can easily accept, and though these happenings apply equally to weeds, the wild ones, they are perhaps easier to appreciate with cultivated plants. For instance, the *Alliums*, members of the onion tribe, are so rich in their sulphur secretions, excretions and 'breath' that they can greatly influence other plants. Garlic, chives and shallots appear to be good neighbours to most garden plants except peas and beans, who hate them and are noticeably retarded when oniony plants are near by. My broad beans are usually excellent, but they failed twice when grown near shallots. Peas and beans prefer carrots and turnips as neighbours and broad beans are happy, too, near early potatoes, and, curiously, they will tolerate leeks. Carrots are so encouraged by the friendly chives and shallots that they grow larger, stronger and more flavorous when near them. Carrots and leeks do well together and the leeks discourage the carrot-fly pest.

Roses are vastly improved in stamina and their scent is doubled by the proximity of anything oniony; they so relish garlic that in some countries where they are commercially cultivated for perfumes, garlic is grown with them. In a private garden this particular companionship may be, aesthetically, difficult to provide, but it pays to give all the onion waste as a mulch for our roses, and we can plant among them ornamental members of the Allium family, such as the golden garlic (*Allium Moly*), which has wide straplike leaves and round clusters of large, brilliant yellow blooms, and the giant chives (*A. Schoenoprasum sibiricum*), with narrow leaves and rose-purple flower-heads. There are more of these decorative onions that we and our roses will enjoy, and in a kitchen emergency their leaves can supply an onion flavour.

I have found that roses grown near any member of the onion family suffer less from black spot than do others that are planted away from this affinity. As this disease is only serious in pure air, and does not occur near industrial districts, I wonder if the sulphurous volatile oil that is characteristic of the Alliums has any such deterrent effect as the sulphur in smoke fumes.

Roses have another friend in parsley; they aid each other in several ways and for many years they have been planted together to their mutual benefit in at least one famous garden, where it has been found that the parsley greatly reduces the roses' aphis attacks. This friendly herb also assists tomatoes and asparagus, which are excellent bedfellows, and the three companions fight each other's soil pests.

The tomato is one of the few cultivated plants with particular root excretions that can vanquish the extremely persistent weed, couch grass. I read of this curious conquest long after I had actually experienced it. During the war I rented half an acre of a field to grow tomatoes for the Marketing Board. It was a mass of couch grass but I dug and manured it and, hopefully, put in thousands of tomato plants. They yielded some tons of fruits, and the couch grass disappeared from their beds, though it remained around the plot. But tomatoes are wonderfully sympathetic to the cabbage tribe. The helpful effect of aromatic herbs on the cabbages is quite remarkable, especially sage, rosemary, mint, thyme and lavender. In fact, aromatic herbs are the most popular of plant associates and should be lavishly included in any garden. They have an invigorating effect on most plants, and a repellent effect on many destructive insects above and below ground. Such pests as the cabbage fly are put off by their scents. Some authorities claim that hyssop is the encouraging herb to plant under a grape vine as it

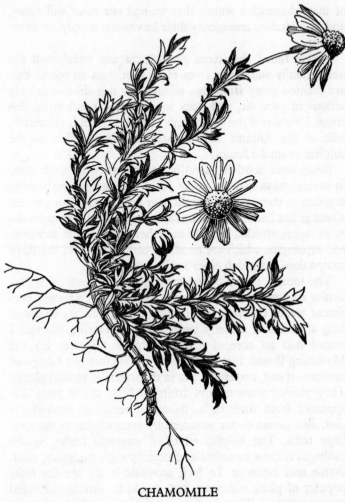

CHAMOMILE

increases the yield and the quality of the fruit. While chamomile assists most things, the cabbages thrive in its vicinity, and as the common chamomile does not mind being trodden down, but breathes its scent more strongly when crushed, it makes a delightful subject for permanent paths between crops in the vegetable plot, or anywhere else in the garden. This herb was once known as the plants' physician, since when it was placed near a sickly one it soon recovered. Among the aromatic herbs, rue is the opponent of sage and sweet basil, for if they are put into the same bed one or all will give up and die.

Lettuce and carrots are mates, as are lettuce and strawberries. Long ago an affinity was experienced by some old gardeners who recorded that strawberries had a curious liking for pine and spruce that was not shared by other plants, and when pine needles were given to them as a mulch, with token twigs and cones placed on their beds, the strawberry plants were more vigorous and their fruits had an excellent flavour. This observation is quite true, and these plants have another sympathetic companion in borage, which gives them an extra fillip. Borage is a helpful plant anywhere in the garden and is a generous source of potassium, calcium and mineral salts in the soil and in the compost.

Those lovely flowering plants, foxgloves, are really a necessary inclusion in any garden as they encourage neighbouring plants while stimulating their growth and endurance. The storage of potatoes and other root vegetables is most successful when they have been grown with foxgloves near by. Indoors, their bloom-spires help to preserve other cut flowers and if some of the tea made from foxglove leaves is put into the water for floral arrangements they will last much longer than when foxglove's influence is lacking. There are many beautiful species of perennial foxgloves that are an asset to any border.

Nasturtiums are another gardener's ally, since the pungent essence they secrete is obnoxious to such plant pests as aphis and white fly, and the excretion from their roots into the surrounding soil not only scares root-lice, but is taken up by other plants so that they too are less attractive to pests. Climbing nasturtiums are often placed around apple trees to

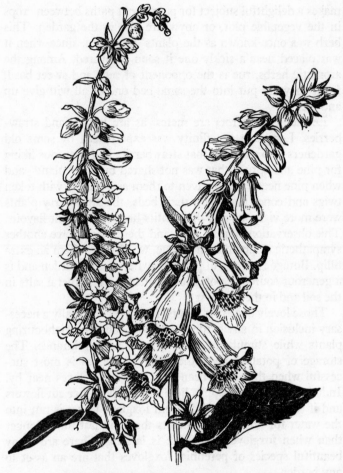

YELLOW PERENNIAL FOXGLOVES

thwart woolly aphis, and they are grown in greenhouses to
frustrate a variety of destructive pests.

The worst villains in any garden are the buttercup tribe,
the *Ranunculaceae*, which give nothing and take all the good
things available, and seriously deplete the ground of

potassium and other elements. The secretions from their roots poison the precious nitrogen bacteria in the soil so that other plants suffer from their deficiency. When the weed varieties, the bulbous or creeping buttercups, or crowfoots as they are often called, get into a strawberry, pea or bean bed, they will dwarf these plants and cause them to panic into producing premature, small fruits. This greedy, selfish family is a large one and its various members such as delphinium, peony, anemone, clematis and many another desirable garden subject, share the same incompatibility with other plants, so that their beds and their neighbours require constant feeding and replenishing.

Another garden favourite of dubious character and breath is the gladiolus. It has been noticed that a bed containing a collection of these showy plants inhibited peas and beans as far away as 50 ft!

It must be obvious that weeds react in the same ways as cultivated plants, secreting, excreting and exhaling their characteristic substances which make them good or bad neighbours, and this important aspect of weeds will be included in their separate descriptions.

A private lawn is the one place in a garden where masses of thriving companionable 'weeds' can be enjoyed, and where their careful assembling succeeds aesthetically and practically. Weed-free turf is now considered necessary for special games, although it is believed Sir Francis Drake rolled his famous bowls on a chamomile green, but weedless turf is a bothersome and highly expensive ambition to strive for, and maintain, in a private garden today. I'm not suggesting that we go back to medieval lawns which were artistic imitations of natural meadows 'starred with a thousand flowers', as we see them in beautiful old tapestries and paintings. These were succeeded by the delightful chamomile lawns, the first of the specialised sowings, which were popular with the purists for several centuries, when labour was plentiful and cheap. In those days grasses would be the hated weeds, but the garden boys would creep along the lawn picking them out. Now that grass has become our ideal lawn plant, and as other weeds inevitably intrude among it, and garden boys have vanished, our most sensible course is surely to encourage, and actually introduce the right weeds. These happen to grow in lawns almost anywhere and they are among the most difficult to control. Using them, we work in Nature's way and with natural help that is free, yet results in a more than ordinarily beautiful, serviceable and scented lawn, which remains green when others are yellowing.

The right weeds are the grass's affinities and its staunchest allies are the clovers. For the tightest growth, white or Dutch clover (*Trifolium repens*), and suckling clover, or lesser yellow trefoil (*T. dubium*), are the best kinds to introduce along with other friendly herbs. These are the bottle-green, feathery-leaved yarrow (*Achillea millefolium*), the dusky-leaved creeping thyme (*Thymus serpyllum*), whose pungent scent was used on the chests of the elegant Greek and Roman males as the perfume of all manly virtues. And there should be the greyish-green foliage of the common chamomile (*Anthemis nobilis*), which smells of apples when crushed. This must not be allowed to flower as it will then become straggly. The colours, textures and the delicious scents of the thyme and chamomile are memorable features of the turf on the Cornish cliffs, also in parts of Surrey and other places in rural England where they grow wild.

With this sympathetic mixture we gain a richly coloured lawn that is permanently and perfectly fertilised. The clovers, besides being a source of sodium, encourage and store the nitrogen-fixing bacteria in their root nodules, so that in the hottest and driest summers a plentiful supply of these elements is available to keep the grass lush and brilliantly green. The yarrow provides copper, nitrates, phosphates and potash; the chamomile gives calcium and the thyme has other gifts. All that is required from the gardener is an annual dressing of the composted mowings, with dried blood and

DECORATIVE ALLIUM

bonemeal to keep the clover in good heart so that it can nourish the grass.

The texture of this turf is thickened to that of a deep, resilient pile, as with the regular mowing, the clover and other plants produce smaller and smaller leaves as they become prostrate and more dense. Unwanted weeds are likely to be smothered, but should buttercups sneak in and be allowed to develop, they will kill the clovers.

It is interesting that this 'weed' mixture, which has produced many fine lawns, was evolved after the 1914–18 War for the Imperial War Graves Commission. They used it all over Europe to make the wonderful, fragrant turf of their huge cemeteries, where it has withstood the varying climatic and soil conditions, and the treading of many thousands of visitors.

When we know that certain weeds are especially retentive of various elements that would benefit our cultivated plants, it is enjoyable, and profitable, to make the weeds into the most natural of liquid fertilisers to be watered on the roots of more deserving subjects. These liquids may also be sprayed on plants to be quickly absorbed by the leaves as folia-feeds, and some of them act as insect repellents at the same time. For this purpose, the young weeds should be gathered in the early morning before they are affected by sunshine. They may be used fresh from the garden or they may be dried in a shady place to be stored in tins for future use.

To make the liquids, cover a handful of the fresh weeds or a tablespoonful of the dried, powdered ones, with a pint of water and bring it just to the boil. Then remove the pan from the heat, keeping it covered as it cools. Strain the liquid free from any bits that would clog the spray, and dilute it with four parts of water. Stir for ten minutes, adding a dessert-spoonful of liquid soap, the washing-up kind, as this helps the stuff to adhere to the foliage when spraying as a folia-feed. The liquid should be used as soon as possible after it is cool and blended. When the infusion is intended to be used only as a root fertiliser, do not add the liquid soap.

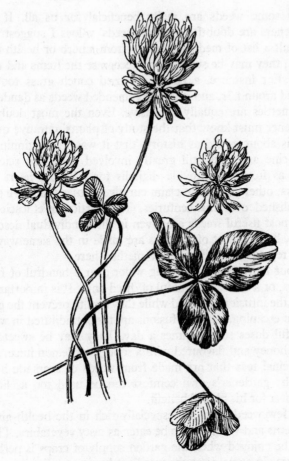

WHITE CLOVER

As our need of elements and plant constituents is very important too, it is equally satisfying to know that some of our weeds can give them to us as well as to our plants. In this aspect, one man's weed can be another man's comforting herb when he appreciates its possibilities. The teas that are infused from some weeds are really beneficial for us all. If any gardeners are doubtful of their weeds' values I suggest they consult a list of medicinal plants from a herb or health food shop; they may be surprised to recognise the items and their cost. For instance, a packet of dried couch grass roots is priced around 3s. and such recommended weeds as dandelion and nettles are equally expensive. Even the most doubting gardener must know that the study of plants' curative qualities is about as old as history, that it was the beginning of medicine and it is still greatly involved with that science. And as foxgloves provide digitalis for relieving heart disorders, other plants can claim curative virtues that have been established over the centuries. The particular efficacies of our most useful weeds are given in their individual descriptions, but as most of the teas are made in the same way, to save repetition I will give the method here.

Pour half a pint of boiling water over a handful of fresh leaves, or a level teaspoonful of dried ones. It is important to keep the infusions covered while they cool to prevent the good steam escaping. These infusions are taken undiluted in wineglassful doses several times a day. They may be sweetened with honey and flavoured with a squeeze of lemon juice. Any medicinal teas that are made from weeds or aromatic herbs for the gardener's own comfort can be used too as liquid fertiliser for his plants' benefit.

A few weeds that are especially rich in the health-giving elements and vitamins can be eaten as tasty vegetables. These may be enjoyed when the garden supply of crops is perhaps meagre. In some cases the weed's value in our diet is greater than that of the usual vegetables we eat. This is certainly true where vegetables have been reared commercially and have absorbed the poisonous pesticides that are now in common use. Some years ago this danger became the problem of American canners of baby foods, after it was found that the vegetables supplied to them contained the lethal components of the pesticides used in their cultivation.

These manufacturers had to insist on the crops being grown in compost and without such chemical aids.

Now, having urged the gardener to take his weeds as medicine, and to eat them as vegetables and salads, I will suggest that he drinks them for pleasure. There are some refreshing and invigorating beers to be brewed from weeds, and there are the wines which can be very good. These are surely appropriate ways to enjoy our weeds most.

The process of making these country wines is dealt with at length in a number of excellent books devoted to the subject. They explain the methods of producing them dry, sweet, still or sparkling, but their instructions are far too long for my space. I can only give the simple process stage by stage hoping it will be easy to follow and that it will produce enjoyable and festive results. The first stage of each brew is given with its recipe in the individual plant's description, but as from the second stage to the final bottling the process is the same for all the wines, the method is given here to save more needless repetition.

Making Wines

The ingredients according to the recipe are put into a large vessel, an earthenware crock is good, or a plastic bucket. This should be capable of holding more than the quantity to be fermented to allow for the frothing. A wooden spoon must be used for stirring, as no metal other than aluminium or stainless steel should ever touch the wine to taint its flavour.

When all the ingredients are in the vessel it must be covered with three or four layers of clean cotton cloth tied down to keep out the tiny vinegar-making flies that will be attracted to the contents. It is left in a warm room for a week or two and should be stirred and squeezed daily, until the exuberant bubbling is over.

STAGE 2. The contents of the vessel are now ladled out and strained through a plastic, aluminium or enamelled colander, into another clean vessel, a plastic bucket or a large bowl being useful. This operation takes out the solid refuse. Then the liquid must be again strained over a large jug through a hair sieve, or the colander with several thicknesses of butter muslin, or one of organdie muslin or flannelette, laid over to trap the bits and the sediment.

The clear liquid is ready to be poured through a non-metal funnel into a glass fermentation jar and lightly sealed with an air-lock. The transparent glass jar is good because you can see how the wine behaves. These gallon fermentation jars are similar to the jars that hold a gallon of fruit squash. Both the jars and the air-locks can be bought quite cheaply from Boots or any store that sells wine-making utensils. A cheap air-lock made of plastic is a cunning device, as it holds a little water to trap the gases from the fermenting wine, but it excludes the air from outside. When dry, an air-lock cork is too big to fix into the neck of a jar, but if it is soaked in boiling water and lightly beaten, it eventually obeys.

When the fermentation jar is filled and sealed it should be kept in a warm room or in an airing cupboard, in as even a temperature as possible, for two or three weeks. Then it is ready to be moved to a cool place for another fortnight or longer, until the little bubbles do not rise to the surface when the jar is moved.

STAGE 3. There will now be a sediment of spent yeast on the bottom of the jar and as this will spoil the wine's flavour if it is not removed, the wine must be siphoned or 'racked' off into a clean jar. This operation requires 4 ft of $\frac{1}{2}$-in diameter rubber tubing, as sold by any chemist. To siphon off, stand the jar containing the wine on a table, and the clean jar on the floor. Insert one end of the tube into the wine above the sediment, suck the other end until the wine flows down the tube, then put this end quickly into the clean jar and let it flow, leaving the brown sediment behind. Insert the air-lock tightly closed and leave the jar in a cool place. As more sediment forms, rack it again, and again if necessary, until the wine remains crystal-clear for a week or two, when it will be ready for bottling. This perfect clearing usually takes about three or four months, sometimes longer, but bottling must not be attempted so long as there is any sign of sediment forming.

STAGE 4. Champagne bottles, or any other good thick wine bottles with an indentation in the base, are the best to use. Some hotels are quite glad to get rid of their empties. Before filling, the bottles must be well washed with detergent and thoroughly rinsed, then dried in a warm oven and cooled with a plug of cotton wool in the tops to keep them sterilised.

To fill the bottles, stand the jar holding the wine on a table, place the bottle on the floor on a tray, insert the tube into the wine, as for racking, suck the end until the wine flows, then put it into the bottle and fill up to about 1 in. from the top.

Use new straight-sided corks and soften them in boiling water. To drive the corks well home it is advisable to invest in a cork-flogger. These gadgets are quite cheap and without one it is nearly impossible to get the cork completely into the bottle neck, whereas with one, a bash with a wooden mallet and the cork is perfectly home. Store the bottles on their sides in a cool dark place, or wrap paper round them to keep out the light. Do not be too anxious to sample the wine as it may not be even palatable for months. Let it mature for about a year, when it should be delicious.

Here follows my little herbal of some familiar and useful 'weeds'—we have corrupted this word and its meaning, it was *wèods*, the Anglo-Saxon name for all herbs or small plants; some they called *wyrt*—our wort. To past generations of men all plants were regarded with respect, some with affection, and some were feared. Many of them were either food or medicine, or they possessed religious or magical influences. A number of the plants we scorn today as our 'weeds' were ready and waiting with their health-giving qualities to serve man and beast long before grasses had numbers of fat ears, or root-crops had thick tubers, or fruit trees produced large juicy fruits. We must owe to them the support of our earliest ancestors and their health and strength to survive and progress. We should never belittle the original and constant value of such herbs.

Archangels

ARCHANGELS. Henbit, red dead-nettle, white dead-nettle and yellow dead-nettle were all included in the popular name Archangel; or they were dead-nettles, because although their leaves rather resembled those of the true nettle they had no stings. But none of them is in any way related to the stinging-nettle, being members of the large family *Labiatae*, whose flowers have pouting lips and whose stems are square. This group of plants includes the favourite enjoyably aromatic herbs such as lavender, rosemary, mint, thyme and many other plants both wild and cultivated.

The henbit so closely resembles the red dead-nettle that it may be thought their points of difference could only concern others of their race or botanising gardeners; but bees are involved, as we are if we would know our weedy friends. Both plants are members of the same genus, *Lamium*, the Latin name coming from a Greek word for throat or gullet, and refers to the specially long tubular corollas of the characteristic, lipped, labiate flowers; and the two plants could be thought of as cousins.

The henbit is *Lamium amplexicaule*, its specifice name meaning stem-embracing, as the upper leaves are stalkless. It prefers a light soil where it can enjoy a long, independent and fruitful life. For although it is an annual, it can be found flowering almost throughout the year. Certainly in mild winters it continues to flaunt its rose-coloured blooms that are mainly for show; there are others, inconspicuous ones, that do not open but self-pollinate the seeds safely inside. So henbit does not rely on bees or any insects, getting along well enough without them, and its average seed production per plant reaches about a thousand, but only a few offspring

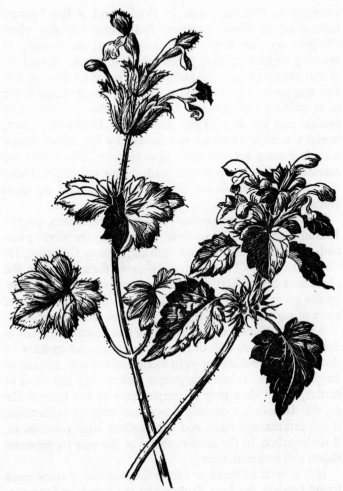

ARCHANGELS: HENBIT AND RED DEAD-NETTLE

survive. Unlike the red dead-nettle, henbit is not gregarious and never appears in abundant masses.

Except in the extreme north, this wild plant ranges all over Europe and down to the Azores and the Canaries; it is common in Palestine, also in Persia, and it has become naturalised in North America. Authorities studying plant affinities say that henbit is so helpful and encouraging to growing vegetables that a few of these plants should be placed among potatoes and other crops in the kitchen garden. For this purpose it is as well to know how to distinguish the more slender and graceful henbit from the sturdier, more prolific and less helpful red dead-nettle. Unlike this plant, henbit's corolla tubes are not hidden by the upper leaves, and they have not the same ring of hairs. The leaves are rounded, rather heart-shaped, with deeply cut edges. Those of the red dead-nettle are tinged with purple and are often clothed with silky hairs.

This plant is *Lamium purpureum*, which means, purple. It is also an annual, "perishing every year; the whole plant hath a strong scent, but not stinking," wrote Culpeper. He also remarked, "To put a gloss upon their practice, the physicians call an herb (which country people vulgarly know by the name of dead nettle) archangel: whether they favour more of superstition or folly, I leave to the judicious reader."

As a wild plant it originated in the mountainous regions of Southern Europe, and developed the energetic capacity of flowering and fruiting for eight months of the year, even in an English garden. It has rose-purple blooms very like those of henbit, but with a ring of purple hairs at the base of the corolla. It does not entirely depend on bees for pollination; it too can manage alone and can produce large colonies, as, if undisturbed, in the colder months of the year its prostrate shoots will root at the joints.

It is of interest that apart from the evidence of some seeds found beneath the Late Bronze Age axe-board, at Stuntney, Cambridgeshire, the introduction of this red dead-nettle and its subsequent naturalisation as a wild plant, a medicinal herb and a common garden weed dates from the Roman occupation of Britain.

The yellow dead-nettle, or yellow archangel, the weazel

ARCHANGELS: WHITE DEAD-NETTLE

snout, was once classed as *Lamium Galeobdolon*, but for structural reasons botanists have now given it a separate genus and it is *Galeobdolon luteum*. This name is not a happy choice, meaning weazel and stench, and condemns this pretty woodland plant to be known all over the world only for the rather unpleasant smell of its crushed leaves and stems, while many a nastier-smelling plant has a nicer botanical name. The attractive flowers are a lovely yellow, which gives the plant its specific name. As a garden weed this perennial archangel is somewhat scarce except in wooded areas and it is quite rare in Scotland. But it does occur in favourable conditions over much of Europe and in Northern Russia.

The white dead-nettle (*Lamium album*) is also a perennial, and a very frequent weed in Britain. Its brittle roots are easily broken to form new plants; and its abundance in hedges, on roadsides and waste places makes it a constant visitor to gardens. The seeds are pollinated by long-tongued, honey-loving insects, mainly bumble-bees whose velvet behinds can be seen wriggling from almost every delightful greenish-white flower, giving this white archangel the popular names bee nettle and honey flower. It is also called Adam and Eve, because if you hold the blossom upside down the black and golden stamens lying side by side are like two sleeping people in a translucent silken tent.

Both this white dead-nettle and its relation the black horehound are among the rare instances of plants imitating another that is quite unrelated for their protection. It is most usual to find them growing and mingling with stinging-nettles, and the plants are so alike in foliage that unless you know your nettles they can deceive you; but the flowers are quite different. They are three natural companions, and their evolved similarity and behaviour is a botanical curiosity.

All the dead-nettles contain a useful supply of elements to enrich the compost for the garden; and they are reputable medicines for the herbalist's practice. Tea infused from them (see page 22) and sweetened with pure honey is a handy domestic remedy in cases of chill. They can also be boiled and eaten as pot-herbs.

To use dead-nettles as a vegetable, gather the young

tender flowering-tips and wash them; then put the wetted leaves into a pan with no more water but with a good slice of butter or margarine. Cook them gently, turning and mixing them occasionally to avoid burning: season lightly with salt and pepper. Strain when tender and serve hot with butter. A tastier mixture is made by adding some chopped chives or spring onions.

Chickweed

CHICKWEED, also known as chick wittles, chicknyweed, clucken wort, skirt buttons, star chickweed. There are quite a lot of chickweeds but the one I am concerned with is the light green, succulent, edible weed that comes up on any good, bared ground. And it must not be confused with the dark green, tough-looking, whiskery mouse-ear chickweed that infests lawns. The eatable star chickweed is smooth and its translucent stems have only a single line of fine hairs that runs up the stems to a joint, then stops, changes sides, to carry on to the next node. It has as glamorous a name as any prima donna; it is *Stellaria media*. The name survives from medieval times as the description of the tiny white star-shaped flowers of its genus: and, specifically, it is media, the middle-sized member of the clan. The chickweeds come of a family with rather splendid members, the *Caryophyllaceae*, which makes this small, modest weed cousin to the bold carnation and the tropical tree whose flower buds are our clove spice.

Chickweed is an overwintering annual seldom without flowers and seeds. This is curious as it has but a thread-like hold in the ground, and its delicate stems and pointed oval leaves appear too fragile to survive any severe weather. It is really a cosmopolitan plant that grows not only throughout Europe and Southern and Central Asia, and has travelled wherever white settlers have gone into temperate lands; but it is established in the Arctic Circle, so that its remains found in the Lea Valley Arctic bed and in the levels of the late-glacial period in Britain are not surprising. Its records are widespread and uninterrupted once this land was cleared of forests and inhabited.

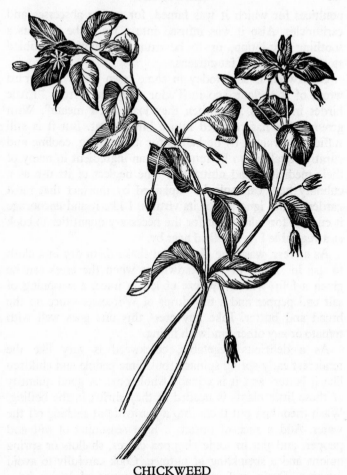

CHICKWEED

This friendly herb is among the few plants possessing a
rich copper content which, with other valuable constituents,
makes it a highly beneficial cress in the diet of man, beast
bird. For cress it is and such a good one that it was once
hawked in city streets and sold in bundles to make salads, to
be cooked as a tender vegetable or to make the effective

poultices for which it was famed, for curing abscesses and carbuncles. Also it was infused into a tea to be used as a soothing eye-lotion, or to be drunk to comfort troubled stomachs or to slim fat citizens.

Chickweed is sold today in some town shops as Gerard wrote of it so long ago in Tudor times, to refresh "Little birdes in cadges . . . when they loath their meate". With groundsel it has relapsed into a birds' treat. But it is still sufficiently valued by herbalists for its soothing, cooling and curative qualities to be retained as an ingredient in many of their medicines and ointments. The neglect of its use as a culinary herb can only be explained by the fact that most gardeners are ignorant of its virtues. I like it and encourage it enough for salads, but for the necessary quantities to cook as a vegetable I go to a field near by.

As a cress, wash the plants and shake them dry in a cloth to use in salads, or for sandwiches when the cress can be given a fillip with a squeeze of lemon juice, a seasoning of salt and pepper and a few drops of Worcester sauce on the bread and butter. Like any cress this one goes well with tomato or any other sandwich filling.

As a delicious vegetable, chickweed is very like the tenderest early spring spinach (but some people and children like it better) and it is equally wholesome. A good quantity of these little plants is needed as they shrink in the boiling. Wash them and put them into a pan without shaking off the water. Add a slice of butter, a light seasoning of salt and pepper, and put in some chopped chives, shallots or spring onions and a sprinkling of nutmeg. Cook carefully to avoid burning. Drain and serve hot. A squeeze of lemon juice before serving gives the vegetable a piquancy that is good with rich meats.

As a tea or tisane, gather a good handful of the plants, wash them and pour over them half a pint of boiling water. The thin yellow peel of a lemon or orange may be added for flavouring. Cover the vessel to prevent the steam escaping.

Drink this amount daily in several doses to relieve constipation, to soothe an upset stomach or to act as a helpful slimming potion of ancient reputation. The same infusion makes an excellent eye-lotion to relieve tired or inflamed eyes.

As an effective poultice, to cure carbuncles or abscesses, place the herbs in a muslin bag and boil for a minute, then apply hot but not scalding. Use the water to bathe the affected part.

The Clovers

THE CLOVERS. White clover, Dutch clover, honeysuckle, honeysuck, bee-bread, bobby-roses, shamrock, three-leaved grass and claver are among the numerous popular names for this fragrant wild plant. Nicholas Culpeper, writing of the clovers early in the sixteenth century, said, "It is so well known, especially by the name of honeysuckles, white and red, that I need not describe them." He continued, "They grow almost everywhere in this land." The white and red clovers are native plants and they range throughout the British Isles and over Europe; they are at home, too, in most of Asia and North Africa; they have been introduced into South Africa, North and South America and into most other lands as crop plants.

These honey-laden bee-plants were the clavers of the Middle Ages until late Tudor times, and they gave their name to places where they grew abundantly such as Clavering, in Essex, Claverdon, in Warwickshire, and Claverton, in Cheshire. The Anglo-Saxons called them *cloeferwort*. Some authorities said the name referred to the trefoil, the three leaflets, and came from the Latin *clava*, a club or cudgel, the three-knotted club of Hercules, the 'club' of our playing-cards. The three leaflets are characteristic of these plants, bringing luck, but when they are found with four leaflets they bring still more luck, and all the leaves are ancient charms against witches and any evil.

Our white clover is called Dutch because some centuries ago it was recognised in Holland as a valuable agricultural crop, but it took British farmers a long time to realise its worth. When they did, they imported great quantities of the seed from Holland.

The round clusters of clover flowers, the bobby-roses, are erect in their young, bee-enticing days, but they droop and fade when their pollination is completed into limp, brown coverings for the developing seed-pods.

When all the grass is burnt with summer's heat, the white clover's leaves remain their beautiful green. This is the plant for clover or mixed lawns, which are lush and verdant in times of drought. The clovers' roots hold tiny sacs of nitrogen to regale the grasses round them so that they too are fed and stimulated. White clover spreads rapidly to form the desirable thick carpet. One little seedling can cover 10 sq ft in a summer's travelling (so noted Curtis in his *Flora Londinensis*).

The botanical name of the clovers is *Trifolium*, meaning three-leaf, and the white clover's specific name is *repens*, which describes its creeping growth (in France it is trèfle rampant). The Trifoliums are members of the same large family as the broom, vetch, sweet pea and other plants with butterfly-like flowers, the *Papilionaceae*, but the clovers' blossoms have but tiny wings.

Red clover, clover-rose, sugar-bosses, honeysuckle, honey-suck, cow-cloos, lady's posies, claver and red meadow clover are names. This purplish-rose-flowered species is *Trifolium pratense*, which means, growing in meadows. Though it is a native plant, it was first introduced into British agriculture in 1645, as marl-grass. Red clover doesn't creep; it grows up to 2 ft tall with several rather lax stems from a single root. Its flowers and leaves are much larger than those of the white clover. Both types have in their roots rich stores of nitrogen with other constituents in their foliage and flowers that are valuable in the compost. Red Clover makes an excellent green manure, grown and dug into the soil, or cut and laid as a weed suppressing mulch, eventually to enrich the ground.

As medicinal plants the clovers have their virtues. They contain sodium, the mineral that reduces acidity, and helps

RED CLOVER

the assimilation of iron in the body, which is sometimes a difficulty; they aid the kidneys and prevent catarrh. The flowers and leaves make a good, gentle tea as a domestic remedy for soothing the nervous system or flatulence, and for relieving bronchial coughs. Clover tea is an old and valued remedy for easing the strain of whooping-cough.

The tea or pleasant tisane is made by pouring 1 pt of boiling water over 1 oz of the plant, flowers and leaves (see page 22).

Both the clovers should be enjoyed as a good country wine. As it is unlikely there will be enough of the honeyed blooms in a garden to make this, it is worth the effort of collecting them from any good clover-growing site.

CLOVER WINE

1 gallon of flowers, red or white or both	*1 gallon water*
3 lemons	*2 oranges*
3 lb best white sugar	*1 oz yeast*

Boil the sugar and water for a few minutes to make a light syrup, then allow it to cool. When it is lukewarm take out a breakfast cupful and crumble in the yeast. Stir it a little and let it work until it makes a creamy liquid. Meantime, put the flowers with the thin yellow peel (no white pith) of the fruits, and the juice, into the fermenting vessel. Pour in the cooled syrup and the creamed yeast. Cover the vessel with a folded cotton cloth and leave it in a warm room for five days. Stir it twice daily with a wooden spoon. Then proceed with stage 2 (see page 23).

Coltsfoot

COLTSFOOT, foalfoot or coughweed, as described in an ancient herbal, "hath many white and long creeping roots, from which rise up naked stalkes about a spanne long, bearing at the top yellow floures, which change into down, and are carried away with the winde; when the stalk and seed is perished, there appeare springing out of the earth many broad leaves, green above, and next the ground of a white, hoarie, or greyish colour. Seldom, or never, shall you find leaves and floures at once, but the floures are past before the leaves come out of the ground". Gardeners will recognise it. The long, creeping roots are, as one old writer said, 'very fat', as fat as a man's finger. Perhaps I am more tolerant of weeds than are most gardeners, as when coltsfoot was in my garden I used to enjoy the patterning of its decorative foliage, like small water-lily leaves made of velvet. But we should all be grateful to this plant, as it chooses to grow in poor, heavy ground, covering with its sea-green leaves infertile land-slips, railway embankments and waste land where no other plant would grow. It will even furnish the cellar floors of dilapidated houses. And its only failure to please is its disappearance in winter. It heralds springtime by dotting the bare ground with its little golden flowers, like miniature, but pale, dandelions. Curiously, the ancient naturalist Pliny did not connect these solitary, shaggy-stalked daisies with the leaves that followed, as they seemed to him to be two different plants. But he did recognise the herb's peculiar curative virtues and recommended his fellow Romans, troubled with obstinate coughs, to burn the dried leaves and roots and swallow the smoke through a reed in the mouth, sipping a little wine after each draught.

42

COLTSFOOT

Nowadays, coltsfoot provides the principal ingredient in herbal tobaccos, and it makes medicines and decoctions for treating any bronchial complaints. It has other curative powers that are exploited in herbal and homeopathic remedies, which can be bought, and coltsfoot candy is very good for coughs. In fact, herbalists and apothecaries so valued this herb that they had it painted as the sign of their healing craft, on their shop doorways.

For a comforting decoction that can be made at home, to be taken frequently when suffering from colds or asthma, boil 1 oz. of coltsfoot leaves in 1 quart of water, until it has boiled down to 1 pt, then sweeten it with pure honey. Coltsfoot tea (see page 22) is good, too, for coughs and bronchitis.

In a garden, coltsfoot eventually yields to good cultivation; hoeing, thick mulches with compost and strawy manure, with dressings of coarse sand to break down the hard, barren ground; peat, leaf-mould and other delicacies can be worked into the soil; for coltsfoot cannot abide in a soft bed, with a banquet. But it can withstand any amount of digging out, as every bit of root that is left behind in the ground, makes a fresh lusty plant, so long as the old, familiar, hungry soil conditions are allowed to persist.

Meantime, while it is still in evidence, coltsfoot's foliage should be composted. It is rich in a number of the soil's requirements, especially sulphur, potassium and calcium, which we can ill afford to waste.

Coltsfoot is of the daisy-flowered order, *Compositae*, and botanists call the plant by Pliny's name for it, *Tussilago farfara*, from *tussis*, cough, and the feminine suffix, *ago*, an obvious reference to its first and continued use. The specific name *farfara* is from Pliny's name for the white poplar, which also has downy leaves. The coltsfoot is downier than most of its daisy relations, and the soft felt, which is easily rubbed off, has had many employments. This was once collected as stuffing for pillows and best beds in parts of Scotland; just as the silky seed-tufts are collected by goldfinches, everywhere, to line their neat cosy nests.

In the days before matches, when tinder-boxes were the lighters, coltsfoot down made the tinder to be sparked aflame by a flint. It was "wrapped in a Rag, and boy'ld in a little Lee", saltpetre was added, then it was dried in the sun.

Coltsfoot wine can be so enjoyable that if the plant is missing from the garden it is worth the seeking in some less pampered site.

COLTSFOOT WINE 1

1 gallon coltsfoot flowers	*1 gallon water*
3 oranges	*1 lemon*
3½ lb best white sugar	*1 oz yeast*

Pick the flowers on a sunny day, shake out any insects that may be lurking among the petals; nip off the stalks as near the head as possible; lay them on a clean tray and leave them for a little while in the sun to dry off any dewy moisture. Then put the flowers into the fermenting vessel with the thin yellow peel (no white pith) of the fruits, and the fruit juice.

Make a syrup by boiling the sugar in the water until it is dissolved, then pour this over the flowers and fruit. When the liquid has cooled to lukewarm, take out a cupful and crumble the yeast into it. When it creams, pour it into the vessel. Leave the vessel covered with a folded cloth, to ferment for seven days in a warm room, stirring daily. Then strain and proceed with stage 2 (see page 23).

COLTSFOOT WINE 2. This recipe takes fewer flowers than the first one, which is an advantage as they are not often easy to find in large quantities. But large, fat, juicy raisins make it a very good wine.

2 qts coltsfoot flowers	*1 gallon water*
½ lb large raisins	*1 lemon*
3 lb best white sugar	*½ oz yeast*

Follow the same method as the first recipe, but chop the raisins before adding them to the other ingredients in the fermenting vessel.

These wines may be ready to sample after maturing for six months, but they will improve if they are kept for a year.

When gathering coltsfoot leaves in the country or from waste ground it is easy to mistake those of the butterbur for coltsfoot's. But butterbur's are large and rounder, whereas the coltsfoot's are heart-shaped with a jimpy edge, with little points, as though they have been bitten all round.

Couch-grass

COUCH-GRASS, also known as twitch-grass, quick-grass, scutch-grass and witch-grass, is a cosmopolitan, at home and flourishing all over Europe and in Northern Asia; it travels around Australia as it does over North and South America. It likes loose ground where it can proceed with its creeping rhizomes near the surface; these underground stems give off side branches with ginger-whiskered bracelets about an inch apart, from which arise new leaf-buds and roots. It advances with pointed, ivory-coloured tips like lances, which force through any obstacles such as potatoes, even tree roots. Couch-grass forms so close and strong a network that—away from the garden—it serves the useful purpose of binding the dunes on sandy seashores almost as efficiently as any other of the special grasses used for this preservation.

The name couch does not refer to the mattress-like webbing of the roots, which could be a good enough reason. It is supposed to have survived from the Anglo-Saxon, *civice*, meaning vivacious, from its own vitality and endurance, which virtues the herbalists of all time have maintained it is able to endow those who use it.

Dog's grass is another name bestowed on it because poorly dogs, and cats, too, seek it to cure their sickness. In fact if it is available they eat it regularly to keep themselves fit and healthy. The seventeenth-century physician, Nicholas Culpeper, after describing this grass, said "If you know it not by this description, watch the dogs when they are sick, and they will quickly lead you to it."

This weed's botanical name, *Agropyron repens*, is less imaginative than the popular ones; it means "field and wheat and creeping". Whatever else you may call it if it is in your

47

COUCH-GRASS

garden, it is a herb of importance. Assessing its medical value, Culpeper said " 'tis a remedy against all diseases coming of stopping, and such are half those that are incident to the body of man; and although a gardener be of another opinion, yet a physician holds half an acre of them to be worth five acres of carrots twice told over".

Twitch-grass is a rich source of potassium, silica, chlorine and other desirable mineral nutrients, with a special carbohydrate and beneficial sugar; and it is distressing to think of the countless fires that have been kindled to burn the long, wiry roots laboriously removed from gardens. Its constituents should make it a boon to any gardener with an ache or twinge, and to his plants. Herbalists sell and use the benevolent sweet-tasting roots to relieve rheumatism, gout, gravel, upset kidneys, catarrhal diseases of the bladder; and they claim that a reasonable course of the tea infused from them is a safe cure for the troublesome infection cystitis. To make the tea, or tisane as it is called in France, infuse 1 oz. of the cleaned and whisker-freed roots cut into short lengths, in 1 pint of boiling water. Cover the jug to prevent the steam escaping. This infusion may be taken freely in wineglassful doses. It is not unpleasant but rather lacking in any particular taste; but it may be flavoured with lemon and a little honey to make it more acceptable; and it is surprisingly effective when taken regularly.

The health-giving qualities of couch-grass apply equally to the garden. The tea gives its virtues to plants as a liquid feed; or the roots supply their stores of nutrients to the compost heap, but only when that is properly made to sufficiently heat up to destroy the roots' chances of survival. Compost containing quantities of couch-grass is especially necessary to replenish the soil where this weed has been colonised; and, curiously, it tends to discourage further outbreaks of the grass when it is applied in generous, thick mulches at least twice in a season.

On the Continent of Europe the couch-grass story is less clouded by the smoke of bonfires than it is in Britain. There it is recognised as a wholesome, health-giving food for horses and cattle, and in Italy the roots are carefully harvested and sold in the markets. In France, too, there is a constant demand for these roots to infuse the popular tisane. In Britain we can buy them from herb shops for two or three shillings a smallish packet, when we have exhausted our garden's stock by using and exploiting the weed to our real advantage.

Daisy

DAISY. Day's Eye, Bairnwort and Bruisewort are among
the popular names for our ubiquitous lawn weed. Daisy is a
corruption of day's eye, the common name in Chaucer's time
as it closed its eyes at nightfall and opened them when day-
light came. Then, Chaucer said he left his bed to see these
flowers, "That blissful sighte softeneth al my sorwe." But
Chaucer was a poet, not a lawn purist. Children made chains
of daisies, so it was Bairnwort, and everybody at some time
made ointment of this plant to heal their wounds and bruises.
It was a great panacea for the crusaders in the Middle Ages
and was then bruisewort. Botanists called the plant *Bellis
perennis* because it was pretty, charming and of perennial
growth. Herbalists have used it in the treatment of varicose
veins, scurvy and other troublous things.

This plant's rosette of spoon-shaped leaves may press
down on the soil to keep its bed free from grassy intruders.
But who could resent a few daisies on a lawn? Certainly not
the poet Clare, who saw "Daisies burn April grass with silver
fires". Nor would St Louis of France, who wore a ring
engraved with the signs of the things he most treasured, a
crucifix for religion, a lily for France and a daisy for his
beloved wife Margaret, who went with him to the sixth
crusade. And the old rose-brick palace at Greenwich bore
everywhere the daisy emblem of Henry VI's Queen Margaret
of Anjou. The flower was the Marguerite in France.

It would be a harsh lawn that did not wear as ornaments
a few frivolous daisies like those lovely, lively old swards of
the Cambridge and Oxford colleges and those of our castles,
cathedrals and much lesser lawns. For curiously, this plant
has always had some close relationship with humans and

DAISY

even yet faithfully marks prehistoric man's tracks across the downs, which are barely distinguishable save from the air.

The daisy is of the *Compositae* family but its seeds lack the usual fluff that would carry them far on the breeze; they fall near the mother plant and some are lugged away by ants while others may be transported by birds. Like its relation Tagetes, this plant has an acrid secretion in its foliage which makes it unpalatable to insects, so that it can provide a gentle pesticide liquid for spraying less lucky plants.

As the mere sight of daisies softened Chaucer's sorrows, so we can be made merrier by the good old country tipple that is made from them. If we can gather enough daisies, flowers, no stalks, to loosely fill a gallon measure, we have the basic ingredient of daisy whisky.

DAISY WHISKY

1 gallon of flowers	*1 gallon of boiling water*
1 lb wheat	*1 lb large raisins*
2 lemons	*2 oranges*
3½ lb best white sugar	*1 oz yeast*

Put the daisies in the fermenting vessel and pour on the boiling water. Let it stand 24 hours then squeeze out the flowers. Add the sugar to the liquid with the chopped raisins, the wheat and the thin yellow rind (no pith) and the juice (no pips) of the fruits. Stir until the sugar is dissolved, then warm to lukewarm a little of the liquid and sprinkle in the yeast. When it froths, add it to the wine and leave the vessel covered with clean cotton cloth, for 21 days, stirring daily. Then proceed with stage 2. This is in the introduction (page 23).

There is an old saying that when you can put your foot on seven daisies summer is come.

Dandelion

DANDELION. This is a naughty weed but an excellent culinary and medicinal herb, with a certain universal fame as a 'potty herb' as many of its popular names suggest—pee-a-bed, wet-a-bed, and in France, Pissenlit. The plant contains elements that healthily stimulate man's whole system, his bloodstream, liver, digestive organs and especially his kidneys and bladder, so if it is taken too greedily the effect could be inconvenient and childish.

As a garden weed, the dandelion, like the nettle, absorbs about three times the amount of iron from the soil taken up by any other plant. It is a miser, too, for copper and for anything else worthwhile in soil nutrients that it can lay its roots to. Above ground, its beautiful, round, flat flowers like heraldic suns provide a rich pollen food and nectar for honey-bees and wild bees necessary for pollinating garden crops. But the plant, like some fruits, exhales a breath that is charged with ethylene gas which hinders the growth of neighbouring plants and gives the same depressing effects that buttercups give, dwarfing plants and causing them to produce premature, pigmy fruits. The thieving dandelion is only acceptable to other plants when it is composted, rotted down, disintegrated to make available its hoard of iron, copper and other things they need. Or it can be made into liquid fertiliser and folia-feed (see page 20) which can help to remedy other plants' deficiencies.

The dandelion also remedies our deficiencies, as it can provide the best source of copper in our diet, along with iron and some other constituents such as taraxacin, inulin, potash, that are contained in its milky juice and have their particular medicinal values. Being non-poisonous, the plant is harmless and entirely beneficial. Though it could not then be scientifically known why it worked, its cures had been experienced for many centuries when they were recorded by Arabian physicians in the tenth century. In England there survives from Tudor times a household book which includes a fascinating list of medicinal, cordial and toilet waters to be regularly infused for the noble's family use, and 'Water of Tantelyon' is among them.

Tea made from dandelion roots or leaves (see page 22) is of great assistance in relieving disorders of a bilious or dropsical nature; it is also a mild aperient; it aids weak digestions and helps to combat anaemia. The dandelion has an ancient reputation for helping to clear gravel, and for dispersing skin eruptions or complexion disorders. A tea made by boiling 2 oz of the root or the leaves in 1 qrt of water, boiled down to 1 pt, is highly effective in cases of eczema, scurvy and all such stubborn skin complaints and this infusion should be taken in wineglassful doses every 3 hours.

This plant's botanical name, *Taraxacum officinale*, arose from medieval Latin from the Persian name meaning bitter potherb, sold in shops. And as an ancient and well-liked culinary herb that can supply us not only with the valuable elements but with more vitamins C and A than almost any other vegetable or fruit, the dandelion should be of great interest to all gardeners for providing a variety of appetising things from those well-nourished, upstanding, juicy specimens flourishing in beds and borders. All parts of the plant have their long-established uses which should be exploited in the kitchen as they used to be. In many an old kitchen garden of palace, manor, rectory and cottage, rows of dandelions bred up to giant size could once be seen, manured and pampered to be served at table, some to be blanched like chicory. When splendid salads were appreciated dandelion's bitterish leaves and shredded or chopped roots with its sweet, tangy flowers were important ingredients.

Today in France these plants are commercially grown to

DANDELION

be sold in markets. The tender young leaves are usually mixed with other vegetables to vary their flavours, or they add piquancy to salads as do the roots. The leaves taste rather like those of Endive.

Salade de pissenlit, a little salad or side salad to be served with a rich meat dish, is made from young and tender dandelion leaves with a light dressing of olive oil and a squeeze of lemon juice. This is garnished with a sprinkling of finely chopped chives, parsley, garlic or borage. This popular French salad is delicious when served, as in their sophisticated restaurants, as *pissenlit au lard*. This is made by trimming blanched pickled pork or bacon, cutting it into small pieces and frying them crisp and dry, then serving them at once, on the raw dandelion salad on a piping hot plate, with a light dressing made of vinegar, a little oil or bacon fat and a seasoning of salt and pepper.

As a vegetable, the dandelion is very good if cooked with care. Wash the tender young leaves and place them in a pan without shaking off the water. Add a large lump of butter, a light seasoning of salt and pepper and cook them slowly until tender, turn the leaves occasionally to mix them all with the butter and to prevent their burning. Strain and serve hot with a squeeze of lemon juice and a sprinkling of chopped chives or parsley.

Dandelion leaves are a happy accompaniment to spinach, but their leaves should be partly cooked first, as they take longer than spinach. Add the wet spinach leaves with a slice of butter when the dandelions are half done. Strain and serve hot with butter and a squeeze of lemon juice.

Dandelion coffee is made from the roots and is so good for us that it is sold in high-class stores and health-food shops and served in vegetarian restaurants. Though this is not quite a fitting substitute for real coffee to be served at a dinner-party, it is a wholesome beverage that tastes like weak coffee but is kinder to delicate stomachs than the real stuff which contains caffeine. This dandelion brew makes a good sleep-inducing nightcap that may be enjoyed by children and invalids. To prepare it, the roots are cleaned and thoroughly dried, then they are slightly roasted to coffee colour in a cool oven. They may be stored for a short time in airtight tins or jars, to be freshly ground when required to make the coffee.

Dandelion beer, apart from being a very popular country

tipple, was the drink most favoured in the past by workers in iron foundries and potteries. It is refreshing and particularly good for relieving stomach upsets or indigestion and for clearing the kidneys and bladder, and it is an enjoyable drink. The whole plants are grubbed up to make it, and the following recipe is worth the making in springtime:

½ *lb young dandelion plants*	*1 gallon water*
1 oz yeast	*1 oz cream of tartar*
1 lemon	½ *oz root ginger*
1 lb demerara sugar	

Wash the plants and remove the hairy roots without breaking the main tap roots. Put them into a pan with the bruised ginger root and the lemon rind (no white pith) and the water, and boil for 10 minutes. Then strain out the solids and pour the liquid over the sugar and cream of tartar in the fermenting vessel, stir until the sugar is dissolved. When the liquid is lukewarm add the yeast and the lemon juice and leave the vessel, covered with a folded cloth, in a warm room for three days. Strain out all the sediment and bottle in screw-topped cider or beer bottles. Store them on their sides. This beer is ready to drink in about a week, when it hisses as the stopper is loosened. It does not keep very long.

A well-made dandelion wine can be very rewarding and as pleasurable as many an expensive foreign one, and it is gratifying to know that while we are enjoying this country brew it is actually good for our health. The wines are fermented from the golden, nectar-rich flower petals. No greenery must come into it as it stops the fermentation and spoils the delicate flavour that should be the wine's attraction. The flowers must be gathered on a dry day and after the dew has left them. The first recipe makes a light, dry sherry-like wine.

DANDELION WINE 1 (Dry)

2 qrt dandelion petals	1 gallon water
3 lb best white sugar	2 oranges
2 lemons	1 oz yeast

Put the cleaned petals into a bowl or plastic bucket that will hold more than a gallon, then pour the gallon of boiling water over them and stir. Cover the vessel and leave for two or three days, stirring daily. Then put the contents into a large preserving pan with the thinly peeled rinds of the fruit (no white pith) and bring to the boil, then simmer for ten minutes. Put the sugar into a fermenting vessel and strain the boiled mixture over it through several thicknesses of muslin or one of cotton or flannelette. When the liquor is lukewarm add the juice of the fruits and stir in the yeast. Leave to ferment in a warm room for two or three weeks, then proceed with stage 2 (see page 23). This wine may be drunk after eight months' maturing but it improves when kept longer.

DANDELION WINE 2 (Sweet)

This recipe makes a sweet wine that must be kept for at least a year before drinking.

3 qrt petals	1 gallon water
3½ lb demerara sugar	½ oz bruised ginger root
½ lb large raisins	1 orange
1 lemon	1 oz yeast

Put the cleaned petals into a bowl or plastic bucket and pour the boiling water over them, cover with a folded cloth and leave for three days stirring several times daily. Strain into a preserving pan with the sugar, the thin yellow peel (no white pith) of the fruits and the ginger root well bashed to

c

bruise it. Bring to the boil and simmer for half an hour, adding more boiling water to make up the original gallon as its evaporates with the boiling. Allow to cool a little, then strain out all the solids through several thicknesses of muslin or one of cotton cloth or flannelette into the fermenting jar. Add the chopped raisins and the fruit juice and when the liquor is lukewarm add the yeast and stir. Cover with a folded cloth and leave to ferment in a warm room for two or three weeks, then proceed with stage 2 (see page 23).

It is always interesting to know how plants came by their popular names and the name dandelion is a corruption of the French *dent de lion*, which is from the Latin name *Dens leonis*. But nobody appears to be sure which part of the plant could be likened to the animal's tooth; some say the golden petals may have been thought to resemble the gilded teeth of the heraldic lion, others prefer the name to be an allusion to the long white tap-root. A more likely explanation seems to have arisen from a surgeon's report in the fifteenth century. He was so impressed with this plant's ability to overcome certain ailments that he said it was as strong and powerful as a lion's tooth.

The dandelion's flowers make it easily recognisable as a member of the daisy tribe, the *Compositae* family, of which many members are notable for their powerful secretions.

Fat Hen

FAT HEN is also known as goosefoot, lamb's quarters, bacon weed, dirty dick, muck hill weed, midden myles, dung weed and melgs, but one should pay no attention to its vulgar names; this is a weed of considerable importance. It is a venerable plant that has but lately passed examination and laboratory tests by an analyst, to prove—or disprove—its claims to the popular name, All Good. And the tests revealed that this All Good contained more iron and protein than either raw cabbage or spinach; it has more vitamin B1 than raw cabbage; and more vitamin B2 than raw cabbage or spinach; and more calcium than raw cabbage: its other constituents compared favourably with both the more popular vegetables.

This common plant persists as a weed, revelling on any farmer's muck heap, gardener's tip, or on pasture; and tolerating almost any waste ground in the British Isles, as it does all over Europe, North and South Africa, America, Asia and Australia. It was once the most valued vegetable for human beings and fodder for their animals. And as its recent analysis has shown, it was a good choice. It lost favour only after its relative, the novel spinach, was introduced from South-west Asia in the sixteenth century. But it left its name in a number of places. It was the Melde of the Anglo-Saxons, growing so profusely in some areas that the settlements were named after it. For instance, the tenth-century Meldeburna —the stream where Melde grew—in Cambridgeshire, which is now called Melbourn. And there was Meldinges, in Suffolk, now Milden. Other places have names as likely to have their roots in this Myles or Melgs or Melde. Its most abundant haunts were too desirable for early food-seekers to pass them by.

62

FAT HEN

We find this plant was growing in Britain in the Late-glacial and the Post-glacial periods. It was in the accustomed diet of the Neolithic, Bronze Age and early Iron Age people; and it was much used by the Romans and later diners.

Fat hen is a rather unlovely annual growing as a spire-shape from 1 to 3 ft tall. It may have reddish streaked stems, or they may be plain green, with short alternate branches. The stalked leaves, too, are variable; some are narrow, some are wide-pointed ovals and toothed, others are almost triangular with wavy teeth. The many clusters of minute, pale green flowers come in short spikes from the axils of the upper leaves. The leaves and stems are powdered with meal of a whitish-grey, especially the undersides of the leaves, and this gives the plant its specific name, album. Its botanical name is *Chenopodium album*, meaning—rather fancifully—goose, and little foot. It belongs to the natural order *Chenopodiaceae*, which also includes some well-known vegetables, the beetroot, sugar beet, spinach and mangold; besides a number of similar wild plants also named goosefoot, with such qualifications as stinking, red, upright, nettle-leaved, fig-leaved and the rest; and they are all so alike that it is difficult to tell which is which. The old favourite vege-table, Good King Henry, *Chenopodium bonus-henricus*, also went out of fashion a century or so ago, but it has returned today, imported and expensive, to be sold in some London shops. Perhaps soon our Dirty Dick will arrive there! Mean-time, it is a weed to be encouraged, even to be grown from seed if it is not already in the garden. It is nutritious and needs no cultivation, but will sow itself to give long supplies of leaves when other greens are scarce. The young seedlings are good in salads, and the tips of the older plants can be taken before they seed; the leaves should be cooked as a vegetable, and they also make good, bright green soup.

In times past, the seeds of fat hen were harvested to be dried and ground into flour for making bread, cakes or gruel, as they are still used in parts of America. They are like buck-wheat in flavour and quite pleasant to eat raw. These were one of the ingredients of the gruel eaten by the Tollund man as his last meal.

There is another good reason for growing this plant, apart from its food value. It is a friendly herb to other plants; its deep roots raise the mineral nutrients in solution from deep in the soil to within reach of its shallower-rooted companions. And its leaf-spread catches and carries water to lesser neighbours. It is a plant to be grown among others as an individual, not as a single crop.

An old English recipe says "Boil Myles in water and chop them in butter and you will have a good dish". But if you use the plant as a vegetable this is too slapdash an instruction for the tastiest results. It is better to wash the leaves and their small stalks in cold water, and put them dripping-wet into a pan with only a spoonful of water to prevent burning. Cook over a low heat and keep moving the leaves. When they are cooked, drain and press out the water. Then chop them finely and return to the pan to reheat, with a light seasoning of salt and pepper and a dash of nutmeg, and a good chunk of butter or margarine. Stir together and serve hot.

For soup, wash a good handful of the leaves and their small stalks and put them into a pan with a pint of cold water, a little salt and a tablespoonful of ground rice. Cook with the lid on until the greenery is tender; then strain the liquid into a basin and rub the solids through a sieve to make the purée. Thin this down to the desired consistency with the liquid and some milk, add a dash of nutmeg and pepper and return to the pan to reheat with a good chunk of butter. Beat with a whip as it simmers until it is blended. It must not boil. When it is sufficiently heated and blended, remove the pan and add a little thick cream or top milk. And to bind the soup in true French manner, pour it into the tureen containing the yolk of an egg beaten with a little of the cooking liquid.

A different flavour can be achieved by cooking with the leaves a few shallots—or a small onion—that have been previously fried in butter.

Ground Elder

GROUND ELDER is also known as goutweed, dog elder, goat's foot, ashweed, bishop's weed, bishop's elder and herb Gerard, and I have written kindly of this herb among my pleasures of wild plants. Now I must put it here among my weeds as it is about the worst of them and the most difficult to control. I have liked it (away from my garden) because it is a herb particularly associated with monasteries, abbeys, castles and the sites of vanished settlements. And being a ruin-fancier I am pleased to find the plants persisting where they were once cultivated and valued. Sometimes remaining there are aromatic herbs and flowering plants to mark old gardens, but always there will be goutweed. It haunts such places that were once busy and are now derelict. But these plants are not ghosts, they are bright green leaves, oval and sharply toothed, on long stalks, that would have smelled and tasted spicily appetising when cooked for dinner in those once peopled refectories.

Goutweed was introduced into Britain in the Middle Ages as a herb of medicinal uses, and a pot-herb. The monastery gardeners received it, and their gouty brothers and out-patients were treated with it, some so successfully that they assumed the herb was the gift of Saint Gerard, who was the particular saint to whom they appealed for relief from this painful disorder. It was then known as the Herb Gerard, being under his special care and influence, and it also became the Bishop's Weed, probably because those dignitaries suffered much from gout and their name gave this curative herb a cachet.

No doubt innocent villagers bought, or were given, the roots to plant in their gardens to have the remedy conveniently at hand. And goutweed was obviously too happy in

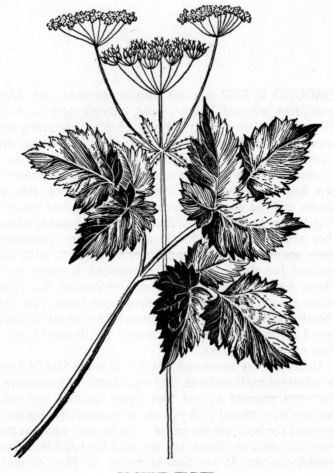

GROUND ELDER

Britain; its rapidly spreading root-stock could, and did, travel; and the dainty umbels of tiny white flowers could produce their flat seed-vessels to be detached and jumped to a distance by the gentlest breeze, so Jack-jump-about it became. And it wore out its welcome in gardens fairly quickly.

By the sixteenth century the famous apothecary and true plant-lover, John Gerard, wrote, "Herbe Gerard groweth of it selfe in gardens without setting or sowing, and is so fruitfull in his increase that where it hath once taken roote, it will hardly be gotten out again, spoiling and getting every yeere more ground, to the annoying of better herbes." That was Gerard the gardener; but as the apothecary he had praise for the plant's effective use for relieving gout. Culpeper was equally enthusiastic for its powers over gout, sciatica and 'jointe-ache'.

Modern herbalists uphold such claims for goutweed's curative properties. It is a kidney-flushing herb and a sedative, working internally when the roots and leaves are taken as a tea (see page 22). And it also works externally through fomentations made from the roots and leaves, boiled together and applied periodically, to the hip for sciatica, and to joints swollen with rheumatic or gouty afflictions. The tea with the poultices often make surprisingly effective cures.

This most persistent weed is an entirely benevolent edible herb, and its leaves make a wholesome and tasty vegetable that is unusually aromatic when cooked. If it is not in the garden (you are in luck) it is worth gathering from the hedge-row, where it is sure to be found.

As a vegetable, wash the leaves in cold water and put them dripping-wet into a pan with a tablespoonful of butter and just enough water to prevent their burning; add a light seasoning of salt and pepper. Cook them gently and keep stirring them about. When tender, strain and serve hot with butter.

This gardener's scourge and aching man's comfort is the botanists' *Aegopodium podagraria*, which means 'goat and little foot' (an obscure reference to the shape of the leaves) and, specifically, 'good for gout', which is understandable. It is of the same large and useful family, *Umbelliferae*, as the carrot, angelica, parsnip and some other very familiar plants with flower-heads like flat umbrellas.

Ground Ivy

THE GROUND IVY, or alehoof, tunhoof, blue-runner, cat's foot, gill-creep-by-the-ground, hedge-maids, Creeping Jenny, was taken by the early settlers to America with their familiar herbs and they used it as they had been accustomed. There it has become Creeping Charlie.

The ground ivy is not as common a garden weed as many other wildings, except in districts near the countryside and where the soil is dampish and not too light and sandy; then it will try to spread its carpet and settle down. As a wild plant it grows everywhere, on waste ground, in hedgerows and on roadside verges; it loves the proximity of oak woods. It flourishes in most parts of Britain and southern Europe, and it creeps over the ground through Asia to Japan. But it is rarely found in northern Scotland, and for its own reasons it will not tolerate the northern Isles, the Hebrides and Orkneys. It is tough enough to be found occasionally in the Arctic Circle; and in Britain it has left its remains in the levels of the full-glacial period in the Lea Valley Arctic bed. So it must have served man for as long as he can remember.

The plant is a perennial with trailing stems and root fibres given off to enter the soil at frequent intervals. By this means it can cover a large area of ground with its evergreen, rather hairy, heart-shaped, scalloped leaves. The lipped flowers arise in little half circles from the leaf-axils, and their colour is variable through shades of bluish-purple. Some blossoms are dark, others are pale, and sometimes there are pure white ones. When growing on a shaded site the leaves are all green, while in a sunny place the stems and foliage will often glow, flushed and stained with crimson and purple.

This carpeting plant's botanical name is now *Glechoma*

68

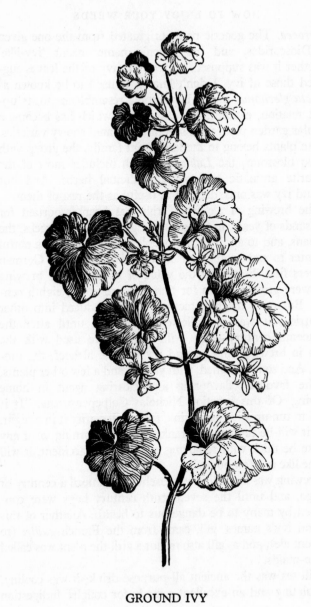

GROUND IVY

hederacea. The generic name originated from the one given by Dioscorides, and the specific name meant 'ivy-like' (whether it was supposed to creep like ivy or the leaves suggested those of ivy, I don't know). It used to be known as *Nepeta glechoma*, and it bears some resemblance to its upright relation, catmint, *Nepeta cataria*, which has become a popular garden plant with several cultivated showy varieties. These plants belong to the large mint family, the group with lipped blossoms, the *Labiatae*, which includes most of the favourite aromatic and many medicinal herbs. And our ground ivy was once deemed as useful as the rest of them.

The brewing of ale from grain has been important for thousands of years, to the earliest Egyptians, the Greeks, the Romans, and to all the other peoples who could boil a potful of water to make some sort of convivial brew. The German brewers first discovered the value of hops in their brewing and were growing them for that purpose in the eighth century. But until the hop method was introduced into other countries—it was not allowed in Britain until after the fourteenth century—only wild herbs were used with the grain in brewing, and ground ivy was the alehoof, the tunhoof. And so it remained, with yarrow and a few other plants, as the favoured flavouring and clearing agent in home brewing. Of this function Nicholas Culpeper wrote, "It is good to tun up with new drink, for it will clarify it in a night, that it will be fitter to be dranke the next morning; or if any drinke be thick with removing or any other accident, it will do the like in a few hours."

Brewing was an ordinary household task until a century or so ago, and until the seventeenth century hops were considered by many to be dangerous to health. Another of this ground ivy's names, gill, came from the French *guiller* (to ferment ale); and as gill also meant a girl, the plant was called hedge-maids.

Gill tea was the ancient all-purpose drink; it was cooling, stimulating and an excellent remedy for coughs, indigestion and kidney complaints. It is still recommended by herbalists and they sell the dried ground ivy. The tea is infused with 1 oz of the herb and 1 pt of boiling water (see page 22). It is taken in wineglassful doses four times a day. This wild plant was once sold in London streets (with its own 'cry') for making curative tea, which was also a lotion for weak, tired or sore eyes; and, with yarrow, it made an effective poultice for abscesses. Being the most effectual herb against pulmonary complaints, it used to be the medicine of hope for consumptives.

Looking down the list of its achievements, from its being "a singular herb for all inward wounds, exulcerated lungs, or other parts", to the curing of gout, sciatica and jaundice, there appear to be few complaints of the body of man, or of beast, that ground ivy cannot cope with; it even cheers away melancholy.

In the garden ground ivy secretes a useful amount of iron, which, with its other constituents, make it a welcome addition to the compost heap.

Groundsel

GROUNDSEL. About groundsel, Nicholas Culpeper wrote, "This herb is Venus's mistress-piece, and is as gallant and universal a medicine for all diseases coming of heat, in what part of the body soever they be as the sun shines upon." He hitched his every herb to a star.

The name groundsel is literally descended from our Anglo-Saxon forefathers, who called the plant *groundswelge*, earth-glutton or ground devourer. In the north of England it is still Grundy-swallow, and it is a greedy weed anywhere and almost everywhere. For it is one of the plants that has accompanied men on their travels and emigrations, and it has settled comfortably in their every colony, save in the tropics, where it found the heat too much, even for groundsel.

Groundsel is an annual of wonderful perseverance; the plants are to be found in flower, and ready to fruit, every month of the year, unless they are hidden under deep snow. No gardener stands a chance of permanently getting the better of it. New seeds will arrive from somewhere, somehow, every day. They sail on the wind and some may settle on wall tops, unnoticed, until it is too late to stop their seeding. Wipe groundsel out with a hoe, and you stir up fresh seeds that within a few weeks will be flourishing plants. On poor soil it rushes into seed-production as a tiny tot; only on good land does it take its time to mature, so that the gardener stands a reasonable chance of cutting it down before the seeds are dispatched. But if he fails, each plant can be responsible for one million descendants in a year.

This weed was once grown as a crop, principally as a food for pigs, goats, rabbits and poultry. Either fresh or dried it

72

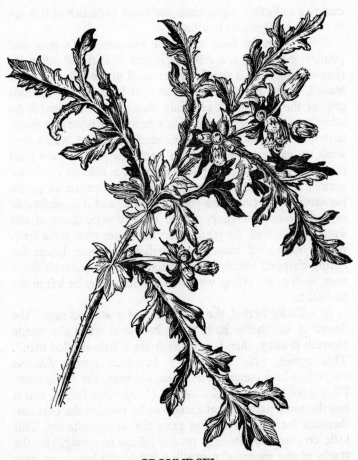

GROUNDSEL

was used in summer and winter, and its effects would be both wholesome and soothing. For the plant contains a sufficient amount of iron and other constituents to remedy animals' debilities and nervous disorders. And it is said, that it will often tempt poorly tame rabbits to take food, and recover. Large quantities of groundsel are sold to be enjoyed by canaries and other caged birds, although too much of it is apt to induce budgerigars to moult.

Apart from its food value for the gardener's pets and poultry, groundsel is a valuable source of iron and other desirable things and it should be returned to the soil. It may be buried, for its life-span is over once it loses its first tenuous grip of the soil; and it certainly should be composted if no other uses are found for the little herb. For several reasons, herbalists still stock this ancient medicinal plant, and a *very* weak tea infused from the fresh weed (see page 22) is a mild purgative and a relief for biliousness; it is also an insurance against scurvy: but a stronger brew is an emetic of gentle persuasion that causes no painful effects, and it is preferable in a domestic emergency to the usual drastic doses of salt water or mustard. By pouring boiling water over some fresh plants, groundsel makes an excellent soothing lotion for curing chapped hands, or roughened skin. This herb's other uses, such as for killing worms in children, may be left to the herbalist.

In a Tudor herbal, the description of groundsel says, "the flower of this herbe hath white hair, and when the winde bloweth it away, then it appeareth like a bald-headed man". This agrees with the plant's botanical name, *Senecio vulgaris*, which comes from *senex*, old man, the Latin name Pliny gave it. Groundsel is of the *Compositae* family, and it has the same particular attraction as its relation the chrysanthemum for the leaf-mining pest, the Marguerite fly. This little creature lays its eggs on the foliage in spring, and the tracks of the maggots' tunnelling inside the leaves are very obvious. The groundsel plants so affected should be buried deep enough to smother the grubs before they emerge as mature flies ready to attack other plants. In this, the weed acts as a sort of useful early trap, and it can be responsible for catching and holding, for a short time, thousands of potential flies.

Horsetail

HORSETAIL, or cat's tail. Nearly two thousand years ago the Roman naturalist, Pliny, gave this odd-looking plant its present botanical name, *Equisetum arvense*, which means horse and bristle, of the fields. He thought its growth resembled a horse's tail but country people in Britain have long preferred to liken it to a cat's tail. This is an almost universal weed that prefers drier ground than is chosen by some of its near relations in the family *Equisetaceae*, who inhabit marshy sites. It is often misnamed mare's tail, which is the popular name for an aquatic plant that lives in the rich beds of lakes. That also has bristly whorls but this water plant is of the genus *Hippuris* and the only representative of the *Hippuridaceae* family. It is in no way related to the horsetails, though at one time the two plants were commonly believed to be males and females of the same species, the girls living in the water, and the boys coming into the garden from neglected fields.

This persistent weed, the bane of many a gardener, resembles a moth-eaten asparagus, with no leaves but scaly sheaths at the joints. It produces much-branched and furrowed green stems which are so hard that they were once used by housewives for scouring churns and pans, and by cabinet makers as fine, sandpaper-like buffers. Horsetail's upper growths are like surfaced periscopes, giving no indication of the industrious bulk of underground ramifications. They spring from hairy, dark brown branching stems going several feet deep into the ground and bearing dark tubers that are ready and very willing to be detached to form new plants. New plants are also sown in spring when unbranched

76

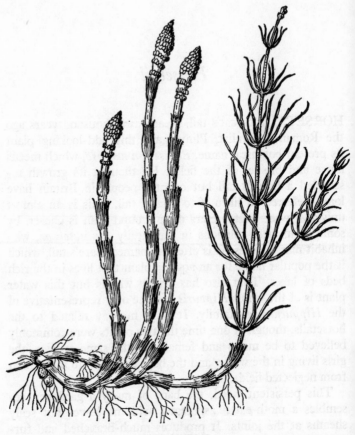

HORSETAIL

brownish shoots with scaly whorls and a cone-like crown appear and scatter their dust-fine spores.

I heard that some workmen building a council estate on a field were puzzled by these masses of roots through which they could hardly dig, nor could they find sufficient top growth to account for them. They did not know about horsetail. But the new tenants would soon learn. However, experiments at Wisley, and those by some famous gardeners who have found this weed established in poor neglected land they have acquired, have shown that it has disappeared after a few seasons of improving cultivation; by generous, deep, enriching mulches, and hoeing to exhaust the weed's efforts as it pokes through the mulch. But bad as it is, horsetail has good uses. It has a great capacity to take up soil nutrients, and anything else it fancies, as was proved many years ago when a Polish physiologist actually recovered a visible amount of gold from the ash of a large crop of horsetail! This unexpected hoard was an early indication of the ability of plants to secrete elements according to their availability, and the plants' particular tastes.

This gold-digger can also secrete very useful quantities of cobalt and calcium, and of silica, which is a particularly effective fungicide for deterring black spot on roses and mildew on any plant. So it provides an effective folia spray to remedy deficiencies and also fungus attacks. While it persists in a garden, it is comforting to know that it can be made to disgorge its properties for the benefit of more splendid and less furtive plants. In fact, horsetail is so good that if it is not in the garden, the dried plant should be bought to make the biodynamic, Preparation 508. This is infused from the stems that arise after the unbranched brownish shoots, bearing the spores, have disappeared in late May. These succeeding branching shoots, like little pine trees, are the richest in the silica and may be used fresh or dried. To make this spray, cover $1\frac{1}{2}$ oz of dried horsetail with 4 qrt of cold water and bring it to the boil. Simmer for 20 minutes only, then take it off the heat and leave it to cool, covered, for 24 hours. Next day, strain the liquid free from bits and use it.

If fresh plants are used, put two good handfuls in a pan and cover with water. Simmer for 20 minutes, then allow it to cool and infuse, covered, for 24 hours. Then strain and dilute with two parts of water to one of the liquid.

Use these infusions once a week and spray the foliage until it drips, then water the roots with this horsetail tea. The infusions are not in any way harmful or poisonous even if they are used in unnecessarily strong proportions of the horsetail liquid to water.

Horsetail tea can be used to prevent mint rust, and the rust that attacks mallows such as hollyhocks.

Dried horsetail is sold as a simple in herb shops, for comforting human beings. The tea (see page 22) is an ancient remedy that has retained the respect of herbalists for its soothing and curative effect on disordered bladders; they use it for bathing inflammations and reducing swellings; it is also prescribed for clearing up ugly skin eruptions, and there are more uses for it to cope with some other complaints.

Potentilla

POTENTILLA. The two potentillas most common as garden weeds are the silverweed and the cinquefoil. Both are lovely as wild plants, and if silverweed were less invasive it would merit a place on the edge of any border for its elegant foliage, as well as for the interest it should arouse. For these potentillas are so closely intermingled with human history that we should surely feel for them some friendly recognition. As their generic name implies, they are potent herbs, members of a group that has such powerful medicinal qualities that their documented uses go back thousands of years to the first Greek doctors; and they remain in the employment of the most modern herbalists.

Silverweed, fern-buttercup, prince's feathers, goose-grass, midsummer silver, silver fern, traveller's ease, are among the charming common names for the *Potentilla anserina*, whose specific name means 'goose', this may have been given because these birds liked to eat its feathery leaves that are silvered and shining with downy white hairs; or perhaps they suggested goose feathers to the people who so named them long ago. These are decorative leaves that could give a fine pattern and contrast in a garden border, just as they enhance floral arrangements indoors, with their long, tapering midrib with pairs of serrated leaflets given off on either side. The flowers are rather flat and of a pale gold colour, but they are too sparse and fleeting to be so useful. In design they resemble strawberry blossoms or miniature wild roses, as they are botanically related to those plants, being of the same natural order, *Rosaceae*.

Silverweed is an almost universal plant that spreads from Lapland to New Zealand, from China to Chile, and it is not

SILVERWEED

faddy about the soil it grows in. It was once valued as a
friendly herb and was cultivated as a crop from prehistoric

until fairly recent times for its roots, which made good eating, tasting rather like parsnips. They were consumed raw, boiled or roasted; or they were dried to be ground into flour for bread and gruel. The plant had medicinal and cosmetic properties, too. Being full of astringent tannin it was used for such purposes as healing ulcers, breaking gallstones, and for beautifying spotty, freckled and sunburned complexions. The leaves, by nature cool, soothed hot, tired feet and were a well-known comfort to foot travellers and those porters and lackeys who trudged the long roads with, or in place of, pack-animals. Footmen, runners or marching soldiers would rest by the wayside and fill up their footwear with these fresh, relieving leaves.

Cinquefoil, five fingers, five-leaved grass, the creeping potentilla, *Potentilla reptans*, the specific name meaning 'creeping', is another attractive garden weed. It has rich green, long-stalked leaves that, in general, are composed of five leaflets, though it may produce more when in our richest soil. The long-stemmed, honey-rich, yellow flowers, resembling those of silverweed, arise from the leaf-axils and they normally have five petals unless the plant flaunts an extra one or two while enjoying a lush diet.

This was a herb of love divinations and of ancient witchcraft; it was an ingredient in 'witches' ointment' that was made from "the fat of children digged out of their graves; of the juices of smallage, wolfbane, and cinquefoil, mingled with the meal of fine wheat". And curiously, it was also credited with such supernatural powers that would keep witches and evil spirits at bay. No doubt these originated from the spiritual significance of the five spreading leaflets. But apart from spells and charms it was a medicinal herb with a high reputation for relieving gout, cancer, throat troubles and other complaints. And, like many another of the familiar remedies of their homelands, it was taken into America by early settlers.

This potentilla's natural range is not so wide as silverweed's but is restricted to Europe and parts of Asia. Both kinds are native plants in Britain and have left their seeds as clues to their early existence. There is one interglacial record of silverweed, but many of late-glacial age in South-east England and Ireland. Then later evidence shows that it was greatly spread by its cultivation and by the clearing of sites. The cinquefoil has also left traces in interglacial deposits but the bulk of its early remains have come from excavated rubbish tips of Roman and later dates, showing the plant's large increase around human settlements where it was cultivated for its varied uses.

These potentillas have similar methods of embarrassing gardeners. They have a blackish, foot-long tap-root that firmly anchors a rosette of leaves. From any leaf-axil comes a slender prostrate branch, a dozen or more from each rosette arranged as spokes of a wheel; and it is no unusual achievement for these runners to travel 6 ft from the parent, touching down every 5 or 6 in. to produce rooted tufts of leaves. Next season the chains of little plants are themselves mature parents, each with a long tap-root and emigrating runners. Thus, one plant can colonise more than 12 sq yds in one season.

The seeds of these lively subjects can come into any garden in manure, or they may be brought by birds since their source of supply is limitless; they grow in waste ground, park, demolition site, hedgerow and verge, quite apart from the countryside. Excluding their tap-roots, both weeds are honest enough to make their network on the surface, unlike some weedy rogues with underground travelling habits. If they are in the garden, or are available in the hedgerow, it is good to know that the whole plants, root and leaf, have the same enriching, curative virtues. With other constituents they are particularly rich in available calcium, and they should be valued for feeding plants or as a simple domestic medicinal aid. The tea infused from them (1 oz to a pint of boiling water) provides a good mouthwash and gargle for ulcerated

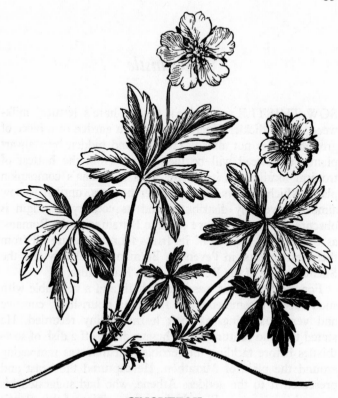

CINQUEFOIL

mouths and relaxed throats. This is also an efficient lotion for clearing the skin of pimples and closing open pores, or for soothing sunburn. An ordinary infusion taken as a tea, with a stronger brew used as a bathing lotion, is recommended for relieving bleeding piles. How true the old apothecaries' claim is of silverweed's ability to fix loose teeth I don't know.

Sow-thistle

SOW-THISTLE, or swine's thistle, hare's lettuce, milk-weed, milky dickle. Wherever there is a garden or a piece of ground that is not already crowded and hidden by stalwart plants, this weed will be there. Only from the hottest of tropical localities is it absent. This is a man's companion plant, which follows him wherever he goes, until it is now almost universally distributed and its country of origin is obscure. It must first have come to Britain with the Romans, as the earliest records of its fruits found here are all from their settlements, in Pevensey, Bermondsey, Ashby and the Silchester site of the Roman town in Hampshire.

For the Romans it was a salad herb and a vegetable with such classic recommendation for being nourishing, curative and very sustaining, as their learned Pliny recorded. He stated that the venturesome Theseus dined off a dish of sow-thistles before tackling the terrible bull that was rampaging around the plain of Marathon. He captured the beast and presented it to the goddess Athene, who had suggested his strengthening menu. Pliny had a great opinion of this plant's medical prowess for relieving such human ills as gravel, bad breath, deafness or wheezing. And nearer to our time, Cul-peper finished his impressive list of its medicinal accomplish-ments with its pleasant use as a cosmetic that "is wonderful good for women to wash their faces with, to clear the skin and give it lustre". To try this claim, the tea should be used as a lotion (see page 22) or the milky juice can be extracted with a juicer.

The sow-thistle is highly valued today as a veterinary herb that is rich in minerals and has such a cooling action that it is advised for the treatment of fevers, high blood-

SOW-THISTLE

pressure, heart disorders and other complaints suffered by animals. And it is instinctively a favourite food among them; often it is taken with eagerness when nothing else will tempt them. One of this plant's ancient names, hare's thistle, and another, hare's lettuce, is said to have been given because, according to one old writer, "when fainting with the heat she recruits her strength with this herb; or if a hare eat of this herb in summer when he is mad, he shall become whole". Older still are the names hare's bush and hare's place, where this hunted animal will rest for safety and peace.

Although the demand for your sow-thistles may be constant to feed the pet rabbits of neighbours who have used up their own supplies, it should be of interest to know that the smoothest, tenderest, young leaves of this ancient pot-herb make a palatable, if rather bitter, addition to salads, as they are eaten on the Continent; or they may be cooked as a vegetable. Not only are their mineral constituents valuable, but they also contain useful amounts of vitamin C.

As a vegetable, wash the tender young leaves and put them into a pan without shaking off the water. Add a good lump of butter or margarine and cook over a low heat. Keep turning the leaves to prevent burning and to mix the butter evenly over them all. Season with salt and pepper to your taste when they are nearly cooked. Like spinach and any other greenery the flavour of this weed is enhanced by the addition of chopped chives or spring onions; and when cooked, a squeeze of lemon juice gives the vegetables an appetising tang. I prefer my edible weeds in a mixture, nettles, dandelions and sow-thistles being a good combination.

The sow-thistle is an overwintering annual, sometimes reaching a height of 4 ft. Its thick, branched stems are hollow and full of milky juice, which in the ancient rule of plants' signatures suggested this one's use for stimulating the yield of milk, and it was given to nursing mothers (human and animal). Of this purpose Culpeper wrote, "The decoction of the leaves and stalks causeth abundance of milk in nurses, and their children to be well-coloured."

The polished leaves are a rich green colour and they are deeply divided into lobes with roughly serrated edges. The yellow daisy-like flowers close to resemble thistle-like heads as these plants are of the same Natural Order, Compositae.

Sow-thistle's botanical name, *Sonchus oleraceus*, is descriptive; the generic name is derived from the Greek word meaning 'hollow', and applies to the stems; the Latin name of the species means 'an edible vegetable'.

This weed's useful performance in the garden is to supply its young leaves, with their vitamins and mineral constituents, to the kitchen and hutch; and to enrich the compost heap with its bulk and goodly stores. When left to its own natural devices it becomes exploited as an hotel for the comfort and convenience of several garden pests. The leaf-miner will lodge in it, lettuce root-aphis spends its holidays in it together with the currant aphides: so that its clean removal as their trap in early summer can liquidate a lot of troublous pests.

Spurges

SPURGES. As enjoyable garden plants with more than ordinary attractions the spurges are more properly called Euphorbias; while their smaller wild relations that invade our gardens as weeds are better known as the spurges. Both names are rooted in their ancient historical associations. The family is a very large group, *Euphorbiaceae*, and the botanical name was retained from the one given long ago by King Juba II of Mauretania, whose wife was Selene, daughter of Antony and Cleopatra. He gave the name Euphorbia to a North African species to honour his physician Euphorbus, who first used this plant medicinally. The common name, spurge, was a corruption of a Latin word meaning 'purge', for which effect the physicians from early times onward employed the plants, though it was extremely drastic and often fatal.

A characteristic of this family is the acrid milky juice which bleeds from the broken stems; as it dries it coagulates in the manner of blood, to seal and heal the plant's wounds. The milk of one exotic member, a tropical forest-tree, becomes elastic as it dries and is one of our sources of india-rubber.

The particular distinction of the spurges is the group of tiny naked flowers, one female accompanied by a number of males sitting amid a cluster of small leaves or bracts. These bracts, giving a petal-like effect as though they were part of a large flower, are especially effective in some of the garden types, where they are brightly coloured either yellow, orange, lime green or spotted; in the weed spurges they are usually of a yellowy-green, but they are still attractive.

A familiar spurge garden weed is the green-flowered petty

WOOD SPURGE

spurge, a small, rather fragile-looking annual growing up to
a foot high with bright yellowy-green, untoothed leaves on
short stalks. It is *Euphorbia peplus*; the specific name was
given by Dioscorides, the Greek physician and botanist, the
rounded bracts encircling the flowers probably reminding
him of a peplus or peplum, the upper robe worn by Greek
women of his day.

The sun spurge, milk-weed, wart grass, devil's milk, mad
woman's milk is another annual that likes gardens, and has

been likened to the spindle of a wooden churn. It branches near the top to carry its broad clusters of toothed, oval leaves and yellow flowers. It, too, has the acrid, corrosive milk, devil's milk, that the curious Tudor, John Gerard, did taste. "Some write by report of others, that it enflameth exceedingly, but my selfe speak by experience; for walking along the sea coast at Lee in Essex, with a gentleman called Mr *Rich*, dwelling in the same towne, I tooke but one drop of it into my mouth; which nevertheslesse did so enflame and swell in my throte that I hardly escaped with my life. And in like case was the gentleman, which caused us to take our horses, and poste for our lives unto the next farme house to drinke some milke to quench the extremite of our heat, which then ceased." This spurge is *Euphorbia helioscopia*, another name from Dioscorides, meaning 'the one who looks at the sun'.

Wood spurge, devil's cup and saucer, Bible leaf, deer's milk, *Euphorbia amygdaloides*, is a perennial that often occurs on the clay lands of Southern England but is rare and local in our northern counties and in Scotland. The reddish stems rise up to 2 ft and the branches are topped by flowering clusters forming a candelabra shape of a beautiful emerald green. The specific name means 'almond-like', because the lower leaves are long and narrow and vaguely suggestive of the almond tree's foliage.

The caper spurge, a biennial, is *Euphorbia lathyrus*. The specific name lathyrus is also the generic name of the sweet pea group, and means 'very impetuous', because one of the peas was used medicinally apparently with violent results; and equally violent are the effects of the caper spurge. This is perhaps a native plant occurring in a few places in English wooded districts and in parts of Wales; but as a garden weed it is most likely to be a survival from garden escapes of long ago, when it was planted in most herb gardens. It is a plant of more than ordinary attractions that place it above any weed but make it a desirable garden subject of architectural

CAPER SPURGE

exactness. Its stiff stems may rise up to 4 ft to carry strong, regularly alternating pairs of opposite leaves of a bluish-green colour, with a peculiar waxy glaze and decorative,

D

whitish midribs. The fruits resemble green capers in appearance, but when used to make caper sauce, as they have been, they can be deadly.

This caper spurge has an old reputation as the mole plant, because it was reputed to be the only thing that really repulsed moles. And it is now confirmed that its root secretions are so effective a mole deterrent that a certain university has been working on the problem of reproducing a synthetic mixture of this compound that will be marketed as a very welcome commodity; it will replace the various methods, which are sometimes effective, recommended to cope with the fur-coated tunnellers. These have included stuffing onions into the runs and dibbling in moth balls or carbide.

The spurges rob the soil of boron, for which the sun spurge is especially greedy, so that these weeds are a valuable addition to the compost heap. And it is said that the spurges actually raise the temperature of the soil around them, making it warm for more tender plants.

Stinging-nettle

STINGING-NETTLE. "Nettles are so well known, that they need no description; they may be found by feeling for them in the darkest night," wrote Nicholas Culpeper in the seventeenth century, to introduce his long praise of the nettle's curative virtues. These were of course confined to the ills of man, long before it was realised that plants were also capable of influencing the welfare of each other. In this context, nettles are the most valuable of weeds. While growing they stimulate the growth of other plants near by and make them more resistant to disease, in the same manner as foxgloves and they, too, improve the storing qualities of root vegetables and of tomatoes.

I had an amazing experience of the effect of nettles on bush fruits. I have six old black currant bushes whose miserable fruits were too small to use. I planted new ones in another place and left the ancient ones uncared for until they were completely hidden by nettles and climbing bellbind. One day I looked into the green domes and found the poor old bushes loaded with large, juicy fruits. No doubt a good nitrogenous feed such as composted nettles or dried blood would have given the invalids a fillip, but I should have missed a very interesting experience.

It is not possible to foster beds of nettles all over the garden, but these weeds should be understood and when once their values are appreciated town gardeners may agree that it would be better to return from a trip into the country with a bag of nettles than with bunches of wild flowers. The nettle's excellent qualities can be as easily enjoyed from harvested plants as from those growing unwanted in our borders. The cut stems and leaves rot down into perfect

STINGING-NETTLE

humus and they can work wonders when laid on the soil under a covering mulch of manure or straw as a weed smother.

These plants thrive on rubbish heaps where the nitrogen bacteria are working on their job of breaking up and decaying vegetable matter. In this process they are a partnership, the nettle encouraging the bacteria while accumulating large quantities of nitrogen, silica and iron, with chlorophyll, protein, phosphates, formic acid and other mineral salts that are required for the well-being of both plants and humans. It is so active a decomposer and humus-maker that decomposed nettles are used to stimulate the fermentation of compost heaps in the biodynamic scheme of gardening, and no heap should lack a generous supply of nettles in its building. The finest rich black humus is to be found in a nettle bed where the stems and leaves have rotted for several seasons. This is the stuff to activate a compost heap, and also to mix with peat when it is used as a mulch, as the nettle humus quickens life into the otherwise sterile peat.

A complete plant food liquid can be made by soaking a sheaf of nettles in a vessel of rainwater for two or three weeks, when the water will contain all the plants' virtues.

A liquid fertiliser made from nettles either fresh or dried, by the usual method (page 20), is not only a good folia feed but also an effective spray against mildew, black fly, aphis and plant lice, in the greenhouse or outside.

The nettle's good points are nowadays often obscured by its stinging ones, but as the venom, which is said to be harmless bicarbonate of ammonia, is dispelled by heat, the plant's medicinal and culinary properties may be as comfortably enjoyed when cooked as they used to be. The nettle can be of great help to remedy deficiencies in the human diet, since it contains most of the elements we require. As a source of iron it exceeds spinach and other recommended vegetables; it can give us our necessary vitamin C, to mitigate our chances of suffering colds or worse debilities. This weed's constituents make it astringent, so that nettle tea is a comforting gargle for sore throats; it is one of the best of antiscorbutics, enriches the blood and relieves bronchial, asthmatic and rheumatic complaints and gout. For its virtues it is in constant use in homeopathic and herbal medicines. Nettle tea is an easily made domestic aid and comfort, but it must not be brewed too strong nor taken too freely or it creates an unpleasant burning sensation. The tea infused from three fresh nettle tops about 4 in long, to 1 pt of boiling water, should be taken diluted with half its amount of water, in wineglassful doses twice daily. This makes a good springtime pick-me-up.

Nettles have always been eaten and drunk; they were so usual a pot-herb and vegetable and the teas, wines and beers brewed from them were so popular that for centuries they were cultivated in most gardens. In *Rob Roy*, Sir Walter Scott described the old gardener at Lochleven raising nettles under cloches as 'early spring kail'. Even though they became coarse and developed irritating crystals in autumn, they still provided a long season of health-giving uses, and they were hawked in city streets to the London cry of "Nettles with tender shoots, to cleanse the blood".

As a spring and summer vegetable, nettles are easily digested and can be very appetising if they are cooked in the following manner:

Creamed Nettles. Wear gloves to gather a quantity of young nettle tops. Wash them and shake off the excess water. Strip off the leaves and put them into a pan with a large lump of butter or margarine. Place the pan on a slow heat and occasionally lift the nettles up from the bottom so that the leaves will all be buttered and equally cooked. As the juices begin to flow, add a light seasoning of salt and pepper. When the leaves are cooked and tender, strain them well and save the juices to make a delicious soup. Reheat the nettles, stirring a little more butter and some cream or top-milk. This vegetable is also very good when cooked with a few chopped chives, shallots or spring onions.

Nettle Soup, using the juice saved from the vegetable dish, is made by combining this with a béchamel sauce. To make this, mix a tablespoonful of butter with a tablespoonful of flour, stirring over a slow heat until it forms a solid roux, then add ½ pt of milk, salt and pepper to taste, and mix well. Then whip it while adding the nettle juice. If a richer soup is liked, grated cheese may be incorporated.

I enjoy nettles in their every aspect, and Samuel Pepys liked nettle 'porridge'. He wrote in his diary on February 25th, 1661, "To Mr Symon's, where we found him abroad, but she, like a good lady, within, and there we did eat some nettle porridge, which was made on purpose to-day for some of their coming, and was very good."

There are a number of recipes for this savoury porridge or pudding, which is still a well-known dish in our northern counties. Many of them tell us to boil the mixture in a cloth in salted water, but this method loses a lot of valuable juices and the pudding is much better and more nourishing when steamed in a covered basin.

Nettle Porridge, or Nettle Pudding, to eat with a hot dinner. Boil until soft a small teacupful of barley to mix with the following ingredients: 6 handfuls of young nettle leaves, 1 handful of dandelion leaves, a small bunch of watercress, a small bunch of sorrel leaves, 8 black currant leaves, a sprig of mint and a spray of thyme, 1 onion.

Wash and chop everything fine and mix with the barley. Season lightly with salt and pepper and add a tablespoonful of butter. Mix all together with a well-beaten egg and put it into a basin, cover and steam for 1½ hours. Serve hot with rich gravy.

This weed can be much enjoyed as a pleasant wine.

Nettle Wine

2 qrt of young nettle tops	*1 gallon water*
4 lb best white sugar	*2 lemons*
½ oz root ginger	*1 oz yeast*

Wash the nettle tops and shake them in a cloth, then simmer them in the water with the root ginger and the thin yellow lemon peel (no white pith) for ¾ hour, adding more water as it evaporates to make up the original gallon. Strain out the solids and add the liquid to the sugar in the fermenting vessel. Stir until the sugar is dissolved. When it is cool to lukewarm, add the yeast. Cover with a folded cloth and leave to ferment for 14 days, then proceed with stage 2 (page 23).

Unlike the wine, which needs to mature for about a year, the traditional convivial nettle beer can be enjoyed within a fortnight of its brewing. It used to be a popular drink for refreshing country workers, and it was especially welcomed to relieve and cheer those who suffered from rheumatic twinges or some other depressing ailments. There are numerous recipes for the beer but the following one is simple and very refreshing. It is very lively and must be contained in strong, screw-stoppered cider or beer bottles.

Nettle Beer

2 lb young cut nettles	1 gallon water
2 lemons	1 lb demerara sugar
1 oz cream of tartar	1 oz yeast

Rinse the nettles and shake them in a cloth, then boil them in the water for about 15 minutes. Strain the liquor into a fermenting vessel over the sugar, the thin yellow peel (no white pith) of the lemons, the lemon juice and the cream of tartar. Stir very well, and when the liquor is lukewarm add the yeast. Keep the vessel covered with a thick folded cloth in a warm room for three days. Then strain out all the sediment and bottle it. The beer will be ready to drink in eight days.

The stinging nettle is an interesting weed with encrusted tails of pale green, male and female flowers that are usually to be found on different plants. This separation of the sexes accounts for its specific name *dioica*, which means 'two dwellings'. The plants have an enjoyable early-morning festival which should be watched, when the males gaily puff their pale gold pollen into the air to be caught by the females. The generic name, *Urtica*, meaning to burn, sting, is as old as Pliny, the Roman who also knew *Urtica dioica's* sting and its great worth. This nettle's flowers may be minute and inconspicuous but as the plant is the only food of the caterpillars of such lovely butterflies as the peacock and the small tortoiseshell, we must be thankful to this weed for our enjoyment of these creatures, its 'flying blossoms'.

The stinging-nettle is a member of a very small family *Urticaceae*, most of which have stings, but curiously, one of its close relations is the innocent little evergreen cress-like plant, Helxine, that we call mind-your-own-business, or mother-of-thousands.

Yarrow

YARROW. Milfoil, yarra grass, thousand leaf, angel flower, bunch of daisies, hemming and sewing, old man's pepper, sneezewort, traveller's ease and woundwort are some of the popular names for this familiar wilding and garden weed. It is a venerable medicinal plant reputed to have been first used by Achilles to staunch his soldiers' wounds, on the instructions of the wise Chiron the Centaur; and in his honour the plant was called *Achillea millefolium*, which means 'thousand leaf', and refers to the numerous divisions of the dark green leaf which give it the attractive feathery or fine fern-like appearance, rather like chamomile. The common name yarrow has survived as a corruption of the Anglo-Saxon *gearwe*. Yarrow's flowers are borne in flat masses and are compound as are all those of the daisy tribe, the *Compositae* family, to which it belongs. It is a perennial with a creeping root-stock that travels underground, throwing up leafy branches to make its compact masses of foliage.

Save in the poorest soils, we find yarrow everywhere, in meadows, pastures, wayside verges and in gardens. It has been introduced into North America and has done well there, as it has in Australia and New Zealand: and in place of hops it makes a headier beer in Africa and Scandinavia.

Yarrow has its value in the garden where it may be encouraged but strictly disciplined. It is one of the desirable plants to be included in a mixed lawn, where it contributes its rich colour and velvety texture and its density of growth, so long as it is mown to curb its exuberance. It is a genial companion plant with root secretions and excretions that organised observations have shown to be helpful and strengthening to its neighbours, while it endows them with

YARROW

more than an ordinary ability to resist disease. Yarrow
multiplies the yield of oils produced by culinary and aromatic

herbs, so that a plant or two set among them helps to intensify their flavours and scents.

As this plant accumulates a rich store of copper and useful amounts of nitrates, phosphates and chlorides of potash and lime, it makes a valuable addition to the compost; and a good liquid fertiliser (see page 20).

The dried yarrow is sold by herbalists as a medicinal herb and it is a good one, with such efficacious constituents and tannin; it also has two peculiar alkaloids, achillein and moschatin, and an aromatic oil. When taken as a tea its effects are stimulating, astringent and healing; it promotes perspiration when that is necessary, particularly at the onset of a feverish cold. It is recommended for kidney disorders; and it is a safe and effective domestic remedy for children's colds, or attacks of measles and other rash-causing complaints. The tea is infused with 1 oz. of the dried herb, or a handful of fresh leaves, to 1 pint of boiling water (see page 22). This may be sweetened with honey and flavoured with lemon. It should be taken warm in wineglassful doses every two or three hours. The same basic infusion is an old-fashioned lotion for preventing baldness, also for healing cuts and wounds. This plant's ancient names, soldier's wound wort, knight's milfoil and herbe militaris, are rooted in its bygone uses when it was carried in the medical supplies for campaigners. As I suffer many cuts and gashes from chisels, saws and cutters, I like the plant's old name, carpenter's grasse, and appreciate that "it is good to rejoyne and soudre wounds" (from the *Grete Herball*, 1526). Enjoyable, too, is yarrow's consequence as a herb so powerful that it could cause or deflect evil—according to the methods used— witches and naughty sprites could make havoc with its help, or it could be used to make trouble for them. So, with the right and good intentions, yarrow was one of the protective herbs that were garlanded about the home, and the church, on Midsummer's Eve to thwart evil spirits at a time when they were most potent.

Weed Control

When a garden is established the best way to keep it as weed-free as possible is to use the weeds as I have suggested to enrich the compost and to benefit the gardener and his family. There is a limit to the number of times even a weed can survive beheading and smothering. Hoeing early will kill the young weeds and mulching the cleared ground generously with a mixture of such materials as compost, manure, leaf-mould, decomposing straw, rotting lawn mowings and peat will smother their successors: any weed that pokes through the mulch can be cut off.

Good regular cultivation and enriching the ground with these humus-forming mulches will soon change the soil's condition and discourage our worst weeds, which will be succeeded by annuals and should present but little difficulty to the gardener with a hoe. It is the nature of weeds to repro-duce in all circumstances. On poor ground this is so urgent that, though starved, they precociously produce seed; often they are such miniatures of their kind that the gardener ignores them. Some weeds can be small as moss yet prolific. But when the soil is in good heart the weeds take their time to grow to maturity and then the gardener has a sporting chance of cutting them off before they can breed battalions of troublesome descendants.

Poisoning these unwanted plants is a bad and expensive practice save in certain circumstances. For instance, to start a new garden on a plot that is choked with the rankest weeds such as docks, thistles, brambles or convolvulus would in-volve some years of patient labour to shift and discourage these subjects. Then a complete killer is useful to make a clean, workable beginning.

Not long ago, apart from the highly dangerous arsenical mixtures, sodium chlorate was the best stuff available for clearing weedy ground before attempting cultivation (and it is still useful for some purposes—slice off a dock and put a tablespoon of the salty stuff on its bared roots). But much time was wasted by its use as it made the ground acid, sterile, infertile, for from six to twelve months, depending on whether the soil was light or heavy. Then it had to be limed and reconditioned.

There is now an almost complete herbicide with none of those old drawbacks. This preparation is made from ammonium sulphamate, a crafty concoction of sulphate of ammonia, and it is only effective in spring or summer when weeds are hungrily feeding. It works through the foliage and the roots, the weeds absorbing it with relish and dying of their greed; a sort of death through over-stimulation. After about a month in the ground the ammonium sulphamate has reverted to sulphate of ammonia, and it acts as that popular nitrogenous fertiliser. The treated area comes to life with a crop of seedlings, mostly annuals, which should be dug in as green manure. In from six to eight weeks the bed, lawn site or new plot is ready for planting or sowing; none of the earthworms, bacteria and other valuable soil inhabitants will have suffered. This treatment was first used by the Forestry Commission to kill such resistant things as scrub, brushwood and the tough rhododendrons infesting their plantation sites. It exterminates most plants, even trees, so that the instructions on the container must be obeyed.

Other complete poisons based on *simazine* were designed to clear weeds from paths and drives. This stuff remains in the ground a few inches deep, keeping it infertile and free from any growth for about a year. These products are rather expensive but as they are the best answer to our time-wasting weedy path problems, they are worth the annual outlay.

Pulling out or forking up deeply rooted weeds damages the fine roots of near-by plants such as peas and beans, and is the cause of roses producing suckers if their roots are accidentally scratched or bruised; such weeds should be cut off with sharp shears or a knife.

But if poisoning is preferred, there are preparations based on paraquat that make hand-weeding among roses unnecessary. As this material destroys the plants' chlorophyll it acts only through their green parts, leaf, stem and shoot; it has no effect on brown bark, so that it may be used with safety in a shrub border, under fruit bushes and between vegetables. It is particularly useful in a rose bed for killing unwanted suckers that spring from grafted stocks. This treatment is best done in spring when weeds and wanted plants in an herbaceous border are just beginning to grow and there is enough space between them to minimise the risk of splashing the wrong target. Paraquat is completely neutralised on contact with soil, so that its effect is harmless but fleeting. As soon as the ground is disturbed in any way, by birds, mice or other animals, or by hoeing or raking after treatment, a fresh crop of weeds appears. This makes it essential that after the successful clean-up of a bed a thick mulch should be laid on as a weed-smother. Apart from annual weeds there are limits to the number of paraquat's victims. Some persistent types like ground elder, convolvulus, horsetail, creeping thistle and other rogues are but little discouraged by one or two applications—couch grass actually recovers to fresh enthusiastic growth! Perhaps when the cost and labour of repeated doses do not count, it would eventually work on most weeds.

In a private garden the value of *selective* weed killers is exceedingly limited. They are not a substitute for good traditional cultivation and must be regarded—if at all—as emergency aids. To employ the selective, so-called hormone, herbicides, you would need a large range of bottles and a great deal of expert advice on their use. They present too great a danger of corrupting distant plants as their deposit is easily carried far by every breeze, and some of them are made volatile in hot weather. I know, because the farmer sprayed his field behind my garden hedge and a week later my shrubs and herbs were badly affected—but the effect on his weeds was negligible! There could also be the danger of build-up in the soil when these little-known concoctions are used.

Some years ago I experimented with one of these chemists' wonder-workers on a bed full of buttercups. The plants went through such paroxysms of distortion and deformity that I was miserable with guilt. But not for long did I—or the buttercups—suffer; they revived after a few weeks. I chopped them up and buried them deep.

I hate poisons but I must confess that soon after I came to this garden I did polish off a crowd of coltsfoot growing in a choice position; I painted the leaves with a brush dipped in a solution of sodium chlorate. And I cleared some nettles from the path into my little wood by putting a pinch of the same stuff, but dry, into each of their growing points. Now I have eaten and composted so many young nettles that there are not many left. After being cut for three years they give up.

While I consider the weed killers I have mentioned are safe when used with discretion, I know that gardeners are being constantly tempted by the apparently unlimited choice of herbicides and pesticides offered for sale in every garden shop. We cannot assess the damage these preparations might do—until it is done—even their makers cannot be sure until the commodities have been in use for a long time. Thalidomide began as a supposed harmless tranquilliser. And there was the case of fluoroacetamide, which was used in the manufacture of a systemic insecticide, 'Tritox'. This was sprayed on food crops by contractors before its dangers were realised. Then through a leak from the factory some ground in Merthyr Tydfil and Smarden was so affected that the cattle grazing on it died. After questions in Parliament and the House of Lords, 2,000 tons of this contaminated soil were sealed in drums and dumped like atomic waste, two and a half miles deep in the Atlantic.

As I write this I am sad and angry. My own garden, usually animated with the flickering patterns of beautiful butterflies, is bereft of them all but two cabbage whites and a small tortoiseshell. I have lost one of my great pleasures. The farmers in this district have had an aeroplane spraying their crops to kill the weed charlock. Apart from the loss of the butterflies for miles around, there are now no skylarks to sing, no ladybirds to graze on the greenfly, and there are very few bees to pollinate the blossoms. What else they have destroyed cannot yet be estimated, but I am sure that in time serious damage will become obvious. When toxic sprays are used to destroy our weeds and enemy pests, they are as surely murdering our friends their predators—and contaminating our food, and the soil's necessary conditioners.

I say stick to the old cultural methods, hoeing and composting, and with respect and understanding use and enjoy your weeds.

The Weeds to Relieve Your Aches, Pains, Boils and Blanes

ABSCESSES .... Chickweed, Ground Ivy and Yarrow
ANAEMIA .... Dandelion, Nettle
ASTHMA .... Coltsfoot, Nettle
BOILS .... Chickweed, Ground Ivy, Yarrow
BILIOUSNESS .... Dandelion, Groundsel
BLADDER DISORDERS .... Couch Grass, Dandelion,
                                                Horsetail, Nettle
BRONCHIAL COMPLAINTS ... Clover, Coltsfoot, Ground Ivy,
                                         Nettle, Sow Thistle
CARBUNCLES .... Chickweed
CATARRH .... Clover, Ground Ivy
CHAPPED OR ROUGH SKIN .... Groundsel, Silverweed
COLDS, CHILLS ...... Archangels, Coltsfoot, Nettle, Yarrow
COMPLEXION DISORDERS .... Dandelion, Sow Thistle
CONSTIPATION .... Chickweed, Dandelion, Groundsel
COUGHS .... Clover, Ground Ivy
CYSTITIS .... Couch Grass
DYSPEPSIA .... Dandelion, Ground Ivy
EYE LOTIONS .... Chickweed, Ground Ivy
ECZEMA .... Dandelion
FEVERISH COLDS .... Yarrow
FLATULENCE .... Clover, Dandelion, Ground Ivy
FOOT DISCOMFORT .... Silverweed
GALLSTONES .... Couch Grass, Silverweed, Sow Thistle
GOUT ... Cinquefoil, Couch Grass, Dandelion, Ground Elder,
                                         Ground Ivy, Nettle
GRAVEL ...... Couch Grass, Dandelion, Nettle, Sow Thistle
HAIR CONDITIONER .... Nettle, Yarrow

INFLAMMATIONS .... Dandelion, Horsetail
JAUNDICE .... Couch Grass, Ground Ivy, Yarrow
KIDNEY DISORDERS ...... Clover, Couch Grass, Dandelion,
Ground Ivy, Yarrow
LIVER UPSETS .... Dandelion
MEASLES .... Yarrow
NERVOUS DISORDERS .... Clover
NETTLERASH .... Nettle
OBESITY .... Chickweed
PULMONARY COMPLAINTS .... Ground Ivy
RHEUMATISM .... Couch Grass, Ground Elder, Nettle
SCIATICA .... Ground Elder, Ground Ivy
SKIN ERUPTIONS .... Dandelion, Horsetail
SUNBURN .... Potentillas
STOMACH UPSET .... Chickweed
THROAT, relaxed .... Potentillas
    „   sore .... Cinquefoil, Nettle
    „   ulcerated .... Potentillas
VARICOSE VEINS .... Daisy
WHOOPING COUGH .... Clover

The teas may be made with more than one of the curative
herbs in a brew. They can be sweetened with honey, or
flavoured with lemon or Marmite (see page 22).

It has been shown recently that many people, both young
and old, suffer from a shortage of vitamin C and the B group
vitamins; some diets lack them to the point of risking scurvy
and repeated cold, with other ailments and loss of vitality.
These essential vitamins could be supplied either by a daily
wineglassful of the juice—squeezed through a juice extractor
—or several doses of the tea infused from such common
weeds as nettle, fat hen, sow thistle, dandelion and chickweed. All together if possible, or as many as can be found.

References

A Modern Herbal. M. Grieve. (Hafner Publishing Co. Ltd., New York.)

Companion Plants. Helen Philbrick & Richard B. Gregg. (Stuart & Watkins, London.)

Flora of the British Isles. A. R. Clapham, T. G. Tutin, E. F. Warburg. (Cambridge University Press.)

Herbal Handbook for Farm and Stable. (Juliette de Baïracli Levy. (Faber & Faber, London.)

The Englishman's Flora. Geoffrey Grigson. (Phoenix House, London.)

The History of the British Flora. H. Godwin. (Cambridge University Press.)

Weeds and Aliens. Sir Edward Salisbury.

The Henry Doubleday Research Association, Bocking, Braintree, Essex. (The Tagetes experiment and other information.)

Glossary of the British Flora. H. Gilbert-Carter. (Cambridge University Press.)

Plant Names Simplified. A. T. Johnson & H. A. Smith. (W. H. & L. Collingridge Ltd., London.)

Index